Nonstandard Notebook

Nonstandard Notebook

Mathematically Ruled Pages for Unruly Thoughts

TIM CHARTIER & AMY LANGVILLE

FOREWORD BY BEN ORLIN

THE UNIVERSITY OF CHICAGO PRESS CHICAGO AND LONDON

The University of Chicago Press, Chicago 60637
The University of Chicago Press, Ltd., London

Published 2024
Printed in Canada

33 32 31 30 29 28 27 26 25 24   1 2 3 4 5

ISBN-13: 978-0-226-83090-2 (paper)
ISBN-13: 978-0-226-83091-9 (e-book)
DOI: https://doi.org/10.7208/chicago/9780226830919.001.0001

Library of Congress Cataloging-in-Publication Data
Names: Chartier, Timothy P., author. | Langville, Amy N., author. | Orlin, Ben, writer of foreword.
Title: Nonstandard notebook : mathematically ruled pages for unruly thoughts / Tim Chartier and Amy Langville ; foreword by Ben Orlin.
Other titles: Mathematically ruled pages for unruly thoughts
Description: Chicago ; London : The University of Chicago Press, 2024.
Identifiers: LCCN 2023057786 | ISBN 9780226830902 (paperback) | ISBN 9780226830919 (ebook)
Subjects: LCSH: Mathematics—Miscellanea. | Mathematics—Popular works. | Notebooks. | LCGFT: Notebooks.
Classification: LCC QA99 .C43 2024 | DDC 510—dc23/eng/20240102
LC record available at https://lccn.loc.gov/2023057786

♾ This paper meets the requirements of ANSI-NISO Z39.48-1992 (Permanence of Paper).

To Tanya and John, for embracing our
nonstandard paths through life

Contents

Foreword

It makes me a little uneasy to write these words, knowing where they'll be printed. After all, the notebook in your hands belongs to you, not me. It's meant to be a channel through which your thoughts can flow, not a bog in which mine can stagnate.

I mean, really. What kind of obnoxious author pens the first entry in someone else's journal?

Then again, we all need something to get us started. If I've learned one thing by writing four books on math, from *Math with Bad Drawings* to *Math for English Majors*, it's that the blank page is not my friend. None of my best ideas are really mine; they're riffs, variations, extrapolations. My thinking is at its freest and most powerful when it's not a monologue, but a dialogue. A conversation.

Which is why this notebook is such a treasure. When choosing minds to engage in conversation, you can hardly do better than those of Amy Langville and Tim Chartier. Both are expert mathematicians. Both are decorated teachers. And beyond that, both are people of delightful gifts and quirky passions.

Tim is a performer: he trained as a mime under Marcel Marceau, and his puppet videos with Bob the Sheep are, for my four-year-old daughter, on par with the Muppets. Meanwhile, Amy has that intimidating, free-floating excellence that I sometimes observe among top-tier athletes. (In high school, she was such a basketball hotshot that the *Baltimore Sun* covered her choice of college.)

So how did this notebook come together?

It began with Amy. In building her interactive Deconstruct Calculus notebooks, she needed to set aside a few blank pages for notes. Inspired by the work of Marc Thomasset (Inspiration Pad) and Matt

Enlow (an art project for the 2019 Joint Mathematics Meetings), she included a few "nonstandard" notebook pages, replacing the familiar lines with parabolas, sine curves, and the like. Her students gobbled them up and demanded more. Could they have a whole notebook in the same vein?

That's when Tim got involved. He became the project's technical director, writing Python code to generate the images as he and Amy brainstormed hundreds of new page concepts.

The result is something special: a startling fusion of three seemingly disconnected purposes.

First, this notebook is a gallery of lovely images. To flip its pages is to take a leisurely stroll through the potentialities of coordinate geometry. When I first saw the book—back when it was just a PDF on a computer screen—I lost myself scrolling. With each new image, I'd say, "Okay, this must be the end. Surely the well of possibilities has run dry." Then I'd advance the page to find that the well, far from going dry, had yielded another brimming bucket.

Second, this notebook is a series of backdoor lessons in mathematics. Its chapters tour landmark ideas in geometry and algebra, from lines to circles to parametric equations. No deep background or intense focus is necessary. Instead, you may be surprised how much you pick up by osmosis (and how curious you are to learn more).

Third, this notebook is, well, a notebook. It's yours to fill however you wish. Write. Color. Doodle. Brainstorm. Compile quotations. Track to-do lists. Record guilty secrets. Or just let each page offer its own unique invitation.

The magic, to me, is that these three purposes coexist on every page. As in an art gallery, each has a title. As in a math lesson, each houses an equation or mathematical concept. But the notebook remains a notebook. Its most important feature is not the title, it's not the equation, and it's certainly not this foreword. What matters most is the space left blank. All the mathematics, all the artistry—it's only there as fodder for your imagination.

So enough of my blather. The pencil is yours.

BEN ORLIN
May 2023
Saint Paul, Minnesota

Introduction

Our civilization has the odd habit of printing books full of lines. Not line drawings. Not lines of poetry or wisdom. Not the boundary lines of a map. Just lines.

Book after book, page after page: faint, horizontal, evenly spaced lines.

We call such notebooks ruled. It's an orderly, lawful word, as befits an orderly, lawful document. The etymology is no coincidence: today, *rule* has many meanings (a binding decree, a reliable pattern, a state of authority), but they all trace back to the Latin *regula*, for a straight stick or ruler. The straight line, you could say, is the first and ultimate rule. The rule from which all others emanate.

Of course, in a ruled notebook, the rules are not the point. What matters is the space between them. The lines are there to invite writing, drawing, and thinking. The rules are only riverbanks, and the river flowing between them is . . . well, whatever you want it to be.

Which raises a few questions: How might changing the lines change the flow of thoughts? What if the straight parallels gave way to curves, clusters, and crisscrosses? What if the once-identical pages began to individuate and develop personalities?

What ideas might come to life, if the rules grew unruly?

. . . .

In a notebook like this one, the typical piece of paper is 5.5 inches wide and 8.5 inches high. It has one vertical line, an inch to the right of the spine, and 26 horizontal lines, 0.28 inches (0.71 cm) apart, beginning from the bottom. We aim to weave those simple elements into new forms. The project rests on a way of translating between numbers and shapes, a mathematical idea called coordinate geometry.

Coordinate geometry begins by naming an origin, a starting point. For now, let's pick the bottom of the spine, where the two pages meet. This is the center of our cosmos, the sun around which all other points orbit.

Now, we can assign each spot on the page a pair of numbers. First is an x coordinate (the spot's distance from the spine); second is a y coordinate (the spot's distance from the bottom). Consider, for example, the top-right corner of the right-hand page. It is 5.5 inches from the spine and 8.5 inches from the bottom. Hence, $x = 5.5$ and $y = 8.5$. In mathematical shorthand, this point is called (5.5, 8.5).

Another example: the point halfway up the spine. It is zero inches from the spine, and 4.25 inches (half of 8.5 inches) from the bottom. Thus, $x = 0$ and $y = 4.25$. In other words: (0, 4.25).

As on a number line, rightward is positive, and leftward is negative. Thus, on a right-hand page (known in book printing as *recto*), all x coordinates are positive. Meanwhile, on a left-hand page (known as *verso*), all x coordinates are negative. For example, the bottom-left corner of a verso page is 5.5 inches left of the spine, and zero inches from the bottom; hence, (−5.5, 0).

Now, let's try a bit of practice. What are the coordinates for the remaining corners of the notebook, and for the center of each page?*

Sometimes it helps to tweak our coordinate system. For example, we might pick a new origin—not the bottom of the spine, but the corner of the page, or the center. We can also change units, from inches to centimeters or millimeters, so that the lines cluster more closely together or spread farther apart. Each system is a different dialect of the same language. In this book, we sometimes switch dialects from page to page, picking the one that makes each image easiest to describe.

"Algebra is but written geometry," French mathematician Sophie Germain once said, "and geometry is but figured algebra." Such is our method. For every spot on the page, a pair of coordinates, and for every pair of coordinates, a spot on the page.

* The top of the spine is (0, 8.5). The bottom of the spine is (0, 0). The lower right of the recto (right-hand) page is (5.5, 0), and the top left of the verso (left-hand) page is (−5.5, 8.5). Finally, the middle of the recto page is (2.75, 4.25), and the middle of the verso page is (−2.75, 4.25).

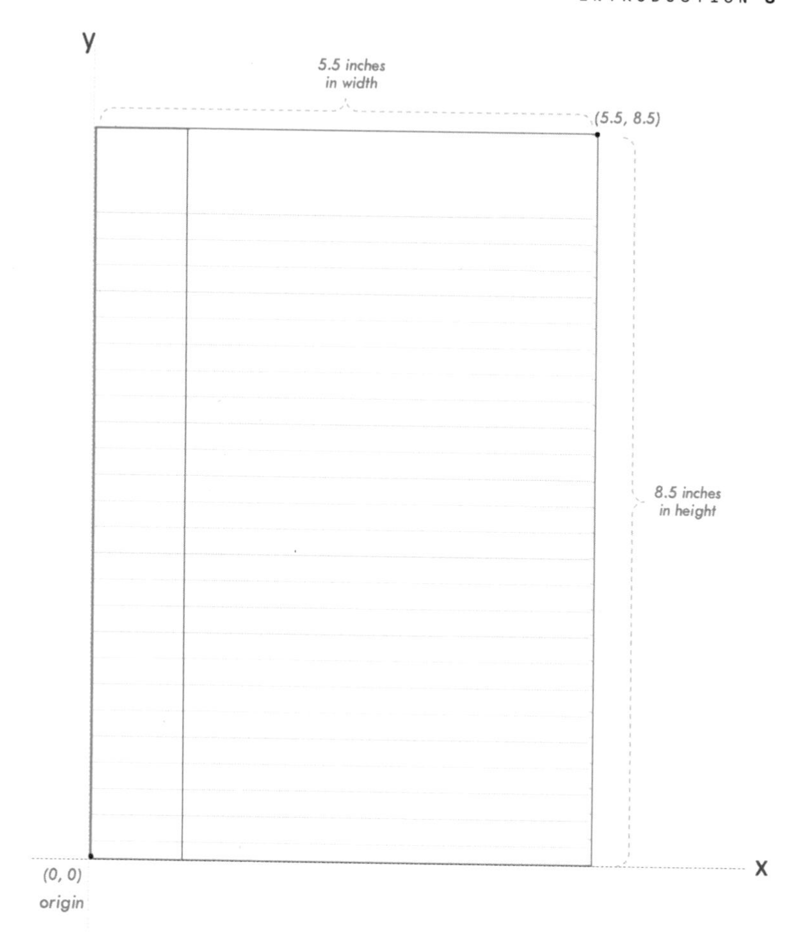

Laying down a coordinate system onto a standard piece of notebook paper

Let that be the first unusual thought recorded in this unusual journal: numbers are only a description of space, and space is only a map of numbers.

....

Then again, a ruled notebook is not a field of isolated dots. It is patterned with lines, and each line is an infinite collection of points. How do mathematicians speak of these?

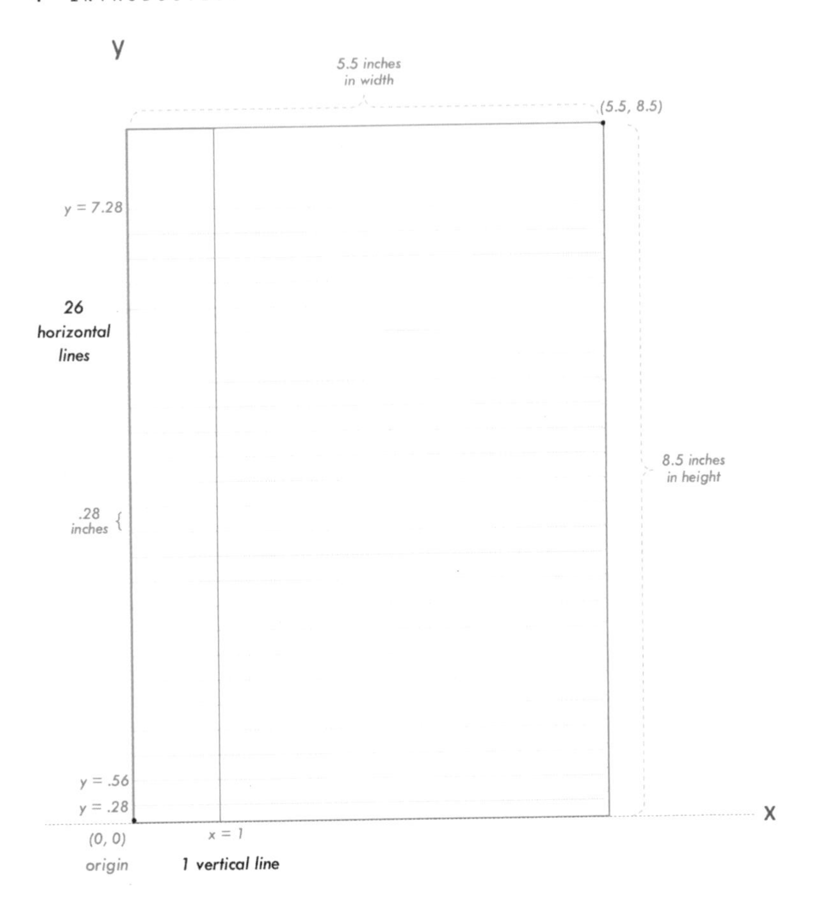

Mathematics of a standard piece of notebook paper

It's best to illustrate with an example. The vertical line on a standard recto page consists of all the points that are one inch to the right of the spine; in other words, the x coordinate must be 1. Meanwhile, the points may be any height on the page, so the y coordinate is unconstrained. Hence, all these points, the entire vertical line, can be described with a single equation: $x = 1$.

What about horizontal lines? Well, the topmost line consists of all the points 7.28 inches above the bottom; hence, the y coordinate must be 7.28. And, since the spots may be any distance from the spine, the x coordinate is of no concern. The line is therefore defined by the shared y coordinate, namely, $y = 7.28$.

Of course, the standard page features 25 other horizontal lines, each requiring a similar description. To list them all can get a bit unwieldy:

$y = 0.28$	$y = 0.56$	$y = 0.84$	$y = 1.12$	$y = 1.40$
$y = 1.68$	$y = 1.96$	$y = 2.24$	$y = 2.52$	$y = 2.80$
$y = 3.08$	$y = 3.36$	$y = 3.64$	$y = 3.92$	$y = 4.20$
$y = 4.48$	$y = 4.76$	$y = 5.04$	$y = 5.32$	$y = 5.60$
$y = 5.88$	$y = 6.16$	$y = 6.44$	$y = 6.72$	$y = 7.00$

So, for the purposes of this book, we employ a mathematical shortcut. Since each of these 26 horizontal lines follows the same form, we refer to the whole group of lines by the single equation $y = c$. The variable c is a parameter, which you can think of as a dial with 26 switches. Set the c dial to 0.28, and you make the bottom line. Set the c dial to 0.56, and you get the second line from the bottom. This continues up to the final setting of 7.28, which yields the topmost line.

With this equipment, we are ready to begin creating new forms. These mathematical rules will allow us to conjure any thought, no matter how unruly.

....

Mathematics does not enjoy a reputation for playful spontaneity. Quite the opposite: it is seen as a realm of structure, a kingdom of suffocating restrictions. If you want to create a nonstandard notebook for unruly thoughts, algebra might seem like the last place you'd turn.

But creativity is not (as we sometimes imagine) a matter of shaking off all constraints. It is about playing against them. We need rules, if only for the sake of breaking them.

And in this sense, mathematics is the first place you should look. Mathematician and author Eugenia Cheng has written that mathematics is not like following a recipe, but "like playing around with ingredients in the kitchen." German mathematician Georg Cantor declared that "the essence of mathematics lies in its freedom." And English mathematician and philosopher Alfred North Whitehead wrote that "the pursuit of mathematics is a divine madness of the human spirit." Our experience bears out these insights. Mathematics is not a box for the mind—or, if it is, then it's only because the mind is like a cat who quite enjoys playing with the box.

We hope this notebook will reveal math's creative side. We will see how straight lines can form fractal crenellations; how circles can disrupt and unify; how waves can create complex landscapes and famous faces. Want to learn more? Try a quick internet search to explore an idea further.

The rules of mathematics, we hope to show, are like the rules of a notebook: invitations to play.

CHAPTER: 1

Lines the wisps of structure

Every art form has its building blocks. Poetry has words; music has notes; chemistry has elements; and mathematics, the language of the cosmos, has the simplest and most elegant building blocks of all.

Straight lines.

A straight line, in the elegant phrasing of ancient Greek mathematician Euclid, is "breadthless width." It is a length with no depth, a kind of structural wisp, only two steps removed from nothingness. In that sense, the lines in this book (being about 0.1 millimeters thick rather than truly breadthless) are not Euclidean lines at all.

But no matter. The line is not a physical building block; it's a conceptual one. A line is a simple idea from which others emerge, an elemental shape from which others are assembled. In this chapter, you'll find lines joining to create sharp cusps, crisscrossing intersections, ornate crenellations, and even the illusion of curvature.

Still, just as every musical chord is composed of individual notes, each of these images is a case study in the symphonic possibilities of the straight line.

classic

$y = c; x = -1$

tilted

$y = mx + c$, where $m = 0$ creates flat lines
and $m \neq 0$ tilts the line; $m = -1/4$ on this page

afloat

$y = c$ for $-4.125 \leq x \leq -1.375$

fire escape

$y = c; y = -0.05x + c$

intersections

$y = 2(x + 3) + c; y = -2(x + 3) + c$

symphony

using the equation $y = c/2$, draw 5 lines, then skip 2 lines and repeat

14 **half empty**

$y = c$ for $0 \leq y \leq 3.9$; $x = -1$ for $3.9 \leq y \leq 8.5$

crux

draw a line from $(0, c)$ to $(2.75, 3.625)$ and another line from $(2.75, 3.625)$ to $(5.5, c)$

convergence

$y = c$ for $-4.25 < x \leq 0$, then connect
the line between $(-4.25, c)$ and $(-5.25, 0)$

herringbone

horizontal lines are $y = c$ shorter each line down and vertical lines are $x = c$ shorter each line across, until you reach the lower-right grid of lines

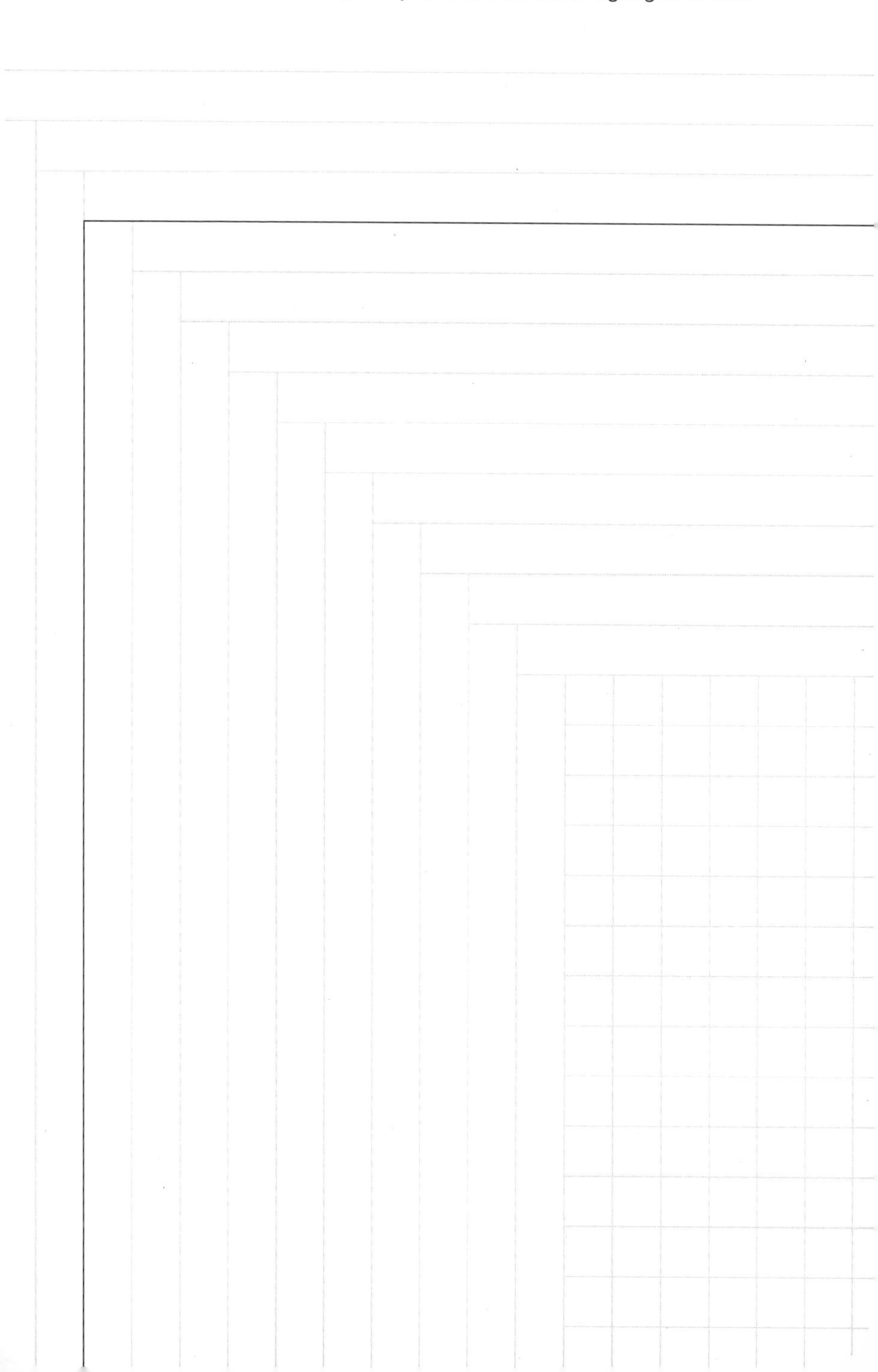

corners

$-x = |y - 4.25| + c$, where $|x|$ is the absolute value function, which leaves positive numbers unchanged (so $|3| = 3$) and turns negative numbers positive (so $|-5| = 5$)

warp speed

$y = c$; connect the points where horizontal lines intersect an oval with the oval's center

heist

$y = c$; create a white diamond formed by connecting the points (0, 3.625), (−2.75, 7.45), (−5.5, 3.625), (−2.75, 0)

divergence

connect points $(0, c)$, $(2.75, 6.37)$, $(5.5, c)$ for c evenly spaced between -3 and 10.25

ladder

$y = c$ for $-3.5 - 0.08(c + 25) \leq x \leq -0.75 - 0.08(c + 25)$

crease

to create the kth line, plot $y = c$ for $0 \leq x \leq 4.95 - 0.14k$;
connect to the line $y = c - 0.05$ for $4.95 - 0.14k + 0.05 \leq x \leq 5.5$

sunfade

$y = c$; break each line into 50 line segments, as you move to the left, make each line segment more transparent

linear curve

draw a line of length 7.25 from (0, *c*) to the bottom edge

skyline

In the technical sense, an *integral* is the area beneath a curve. But in a broader sense—one that has inspired novelists, philosophers, and poets—an integral is a sweeping totality: the sum of infinitely many pieces, each infinitely small. How do you calculate such a curving, flowing thing? Surprisingly, with straight pieces. You create a kind of skyline of rectangles (known as a Riemann sum) that hugs the curve's contours. The skinnier and more numerous the rectangles, the more precise the calculation. Thus, a subtle curvature is subdued by simple verticals and horizontals—lots and lots and lots of them.

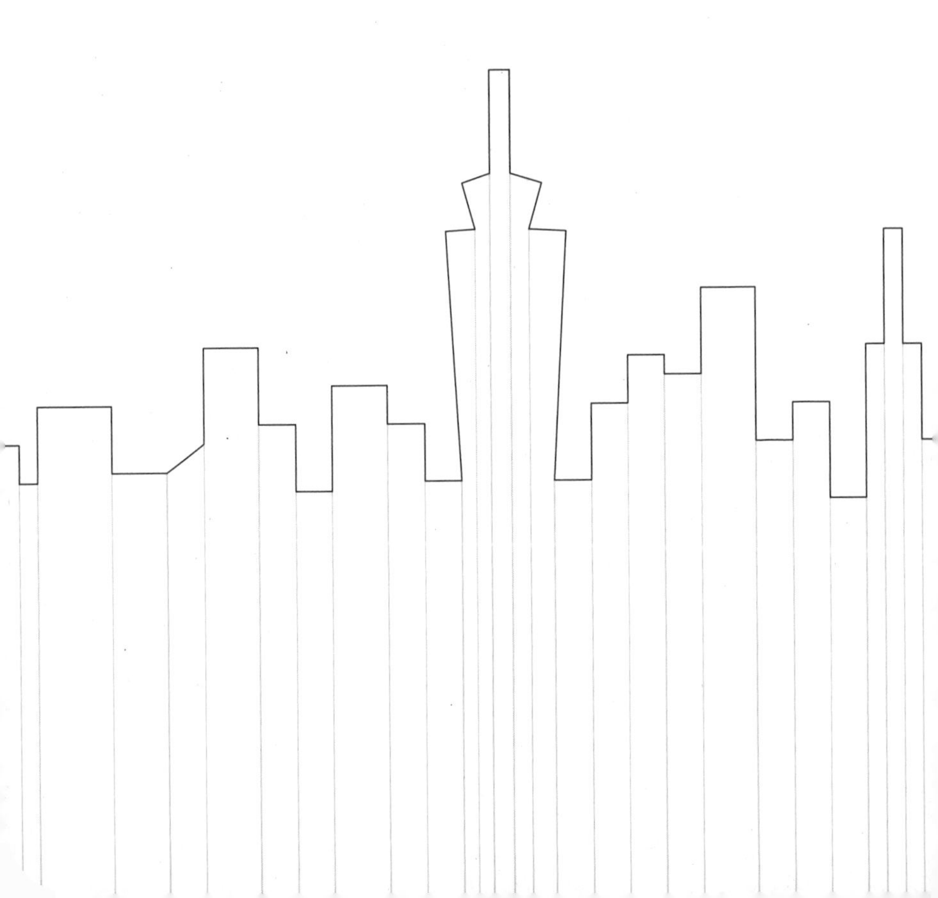

CHAPTER: 2

Parabolas residues of motion

What is the parabola? It is, in so many words, the symmetric U-shaped curve created by slicing a cone parallel to its side. But note that we say "the" parabola, not "a" parabola. This is deliberate.

There is only one parabola.

Hard to believe, we know. After all, this chapter seems to exhibit hundreds of parabolas: smiles, frowns, hills, valleys, rainbows, curtains. Some look sharp and narrow, with hairpin turns. Others look wide and sloping, with gentle crests and troughs. Surely this diversity of appearances speaks to a diversity of geometries?

Not at all. Just as all circles are the same basic shape, writ larger or smaller, so too are all these parabolas merely zoomed-in or zoomed-out visions of the one primordial parabola. Zoom in, and it looks flatter; zoom out, and it looks steeper; but the parabola remains the same.

Why does the parabola matter? For millennia, it didn't. It was a geometric curiosity, an intellectual plaything. But then it was revealed as the shape of the equation $y = x^2$. Anytime a number is squared, the parabola lurks in the shadows. This small fact has big consequences. A thrown rock follows the parabola through a vacuum. A comet traces the parabola through the cosmos. Sir Isaac Newton found this latter fact so striking that, when writing the *Principia*, his masterwork of modern physics, he saved the proof of these parabolic paths for his grand finale.

That said, parabolas are not our finale. Beyond the quadratic x^2 is the cubic x^3, the quartic x^4, and beyond. These curves are more versatile and varied than the parabola—and they, too, will appear in the pages ahead, as variations on the theme. Still, our star performer shall be the parabola, a single actor who can play many roles.

glacial valleys

$y = x^2 + c$, shifted to the left

reflecting pool

$y = (x + c)(x - c)$ for $y < 0$;
$y = -(x + c)(x - c)$ for $y > 0$;
origin shifted to the center of the page

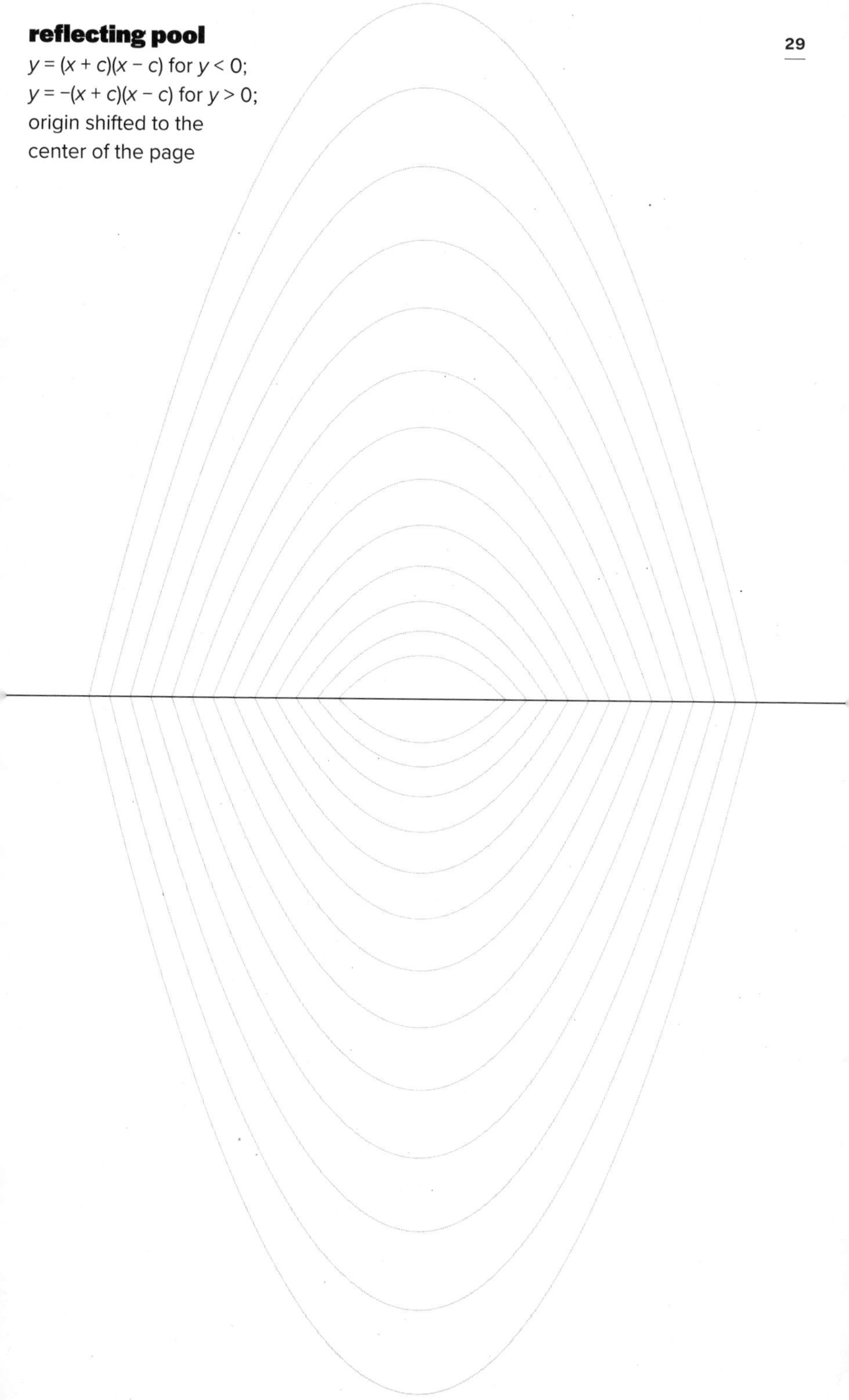

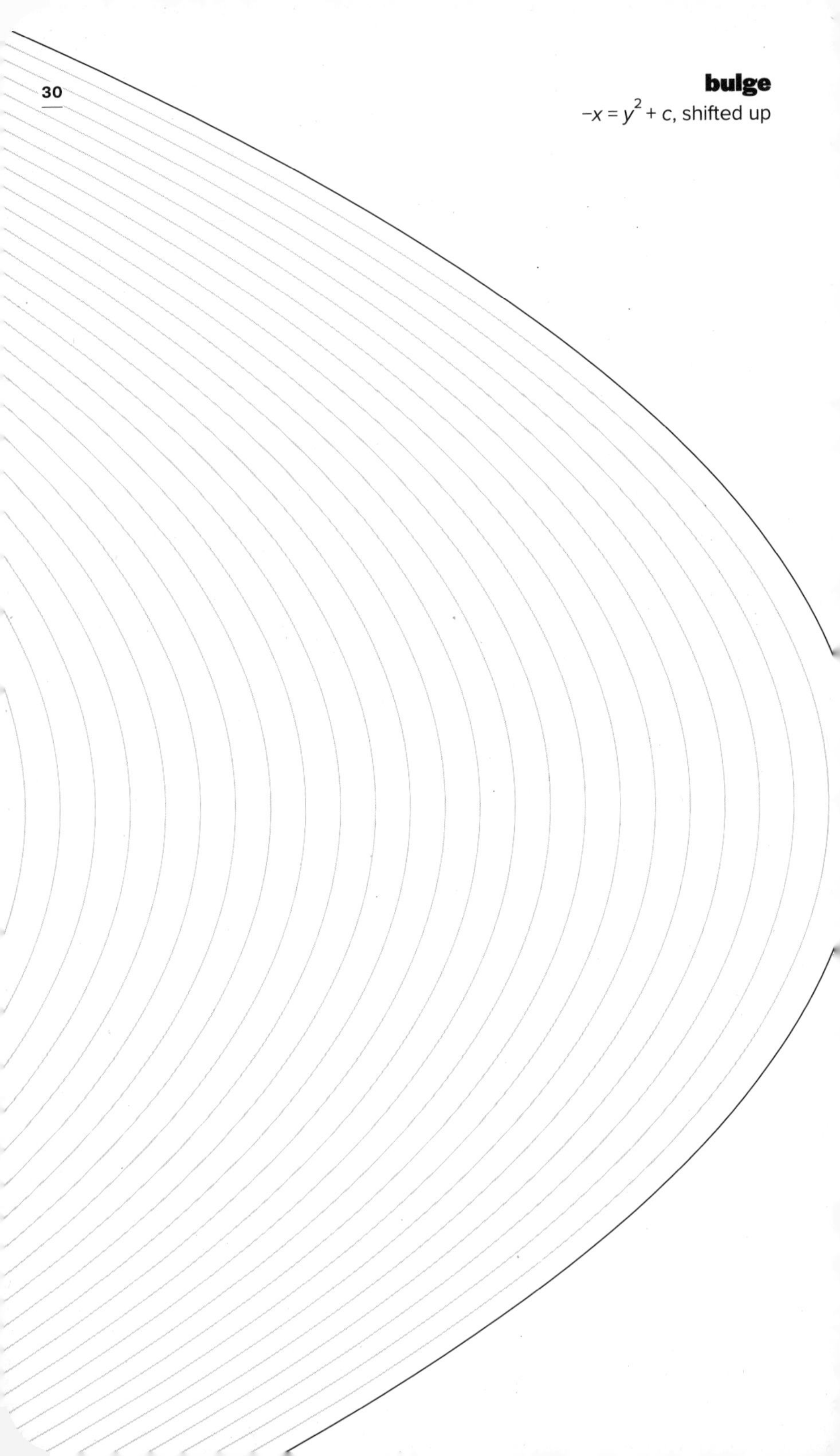
bulge
$-x = y^2 + c$, shifted up

pinch

$y = x^2$ and $y = -x^2$, shifted to the right; if a line would intersect a parabola, draw a line tangent to the curve, otherwise draw a horizontal line

unfurling

$$y = a(x + 2.75)^2 + c$$

for $0 < a \leq 2$

sail

$x = -(y + c/4)^2 + c$, shifted up

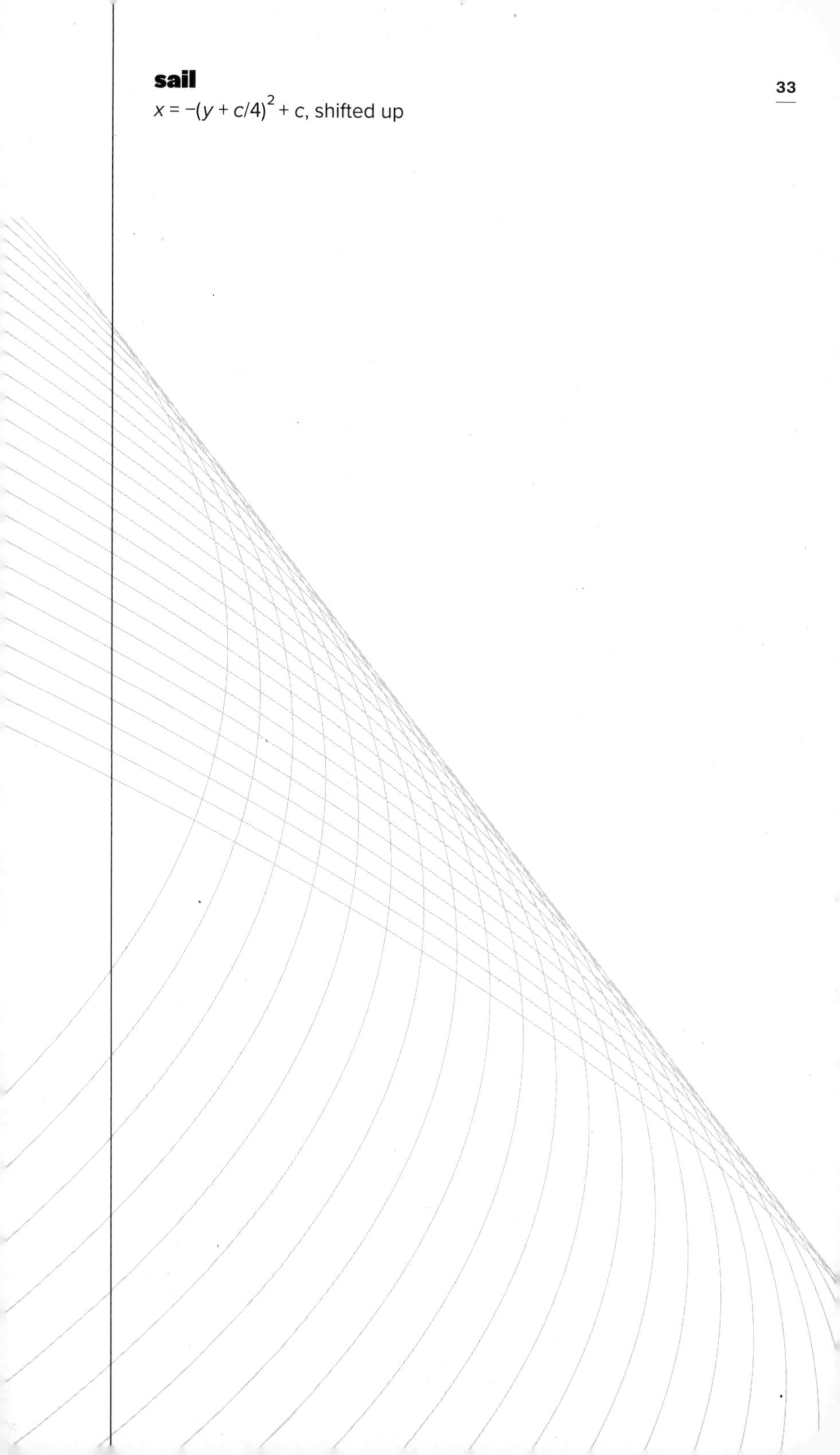

orbits

$y = x^2 + c$ and $y = -x^2 - c$; $x = y^2 + c$ and $x = -y^2 - c$;
origin shifted to the center of the page

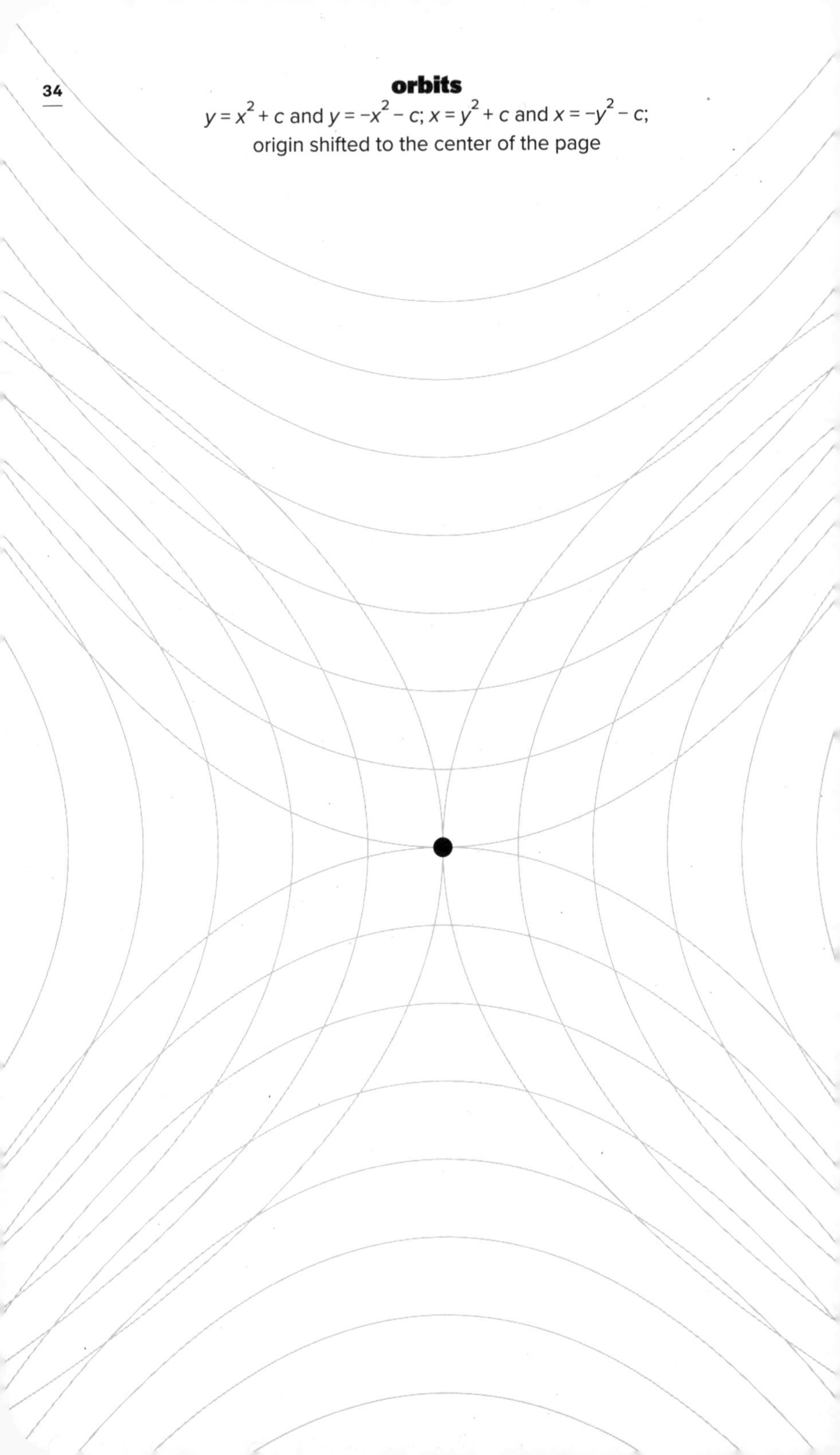

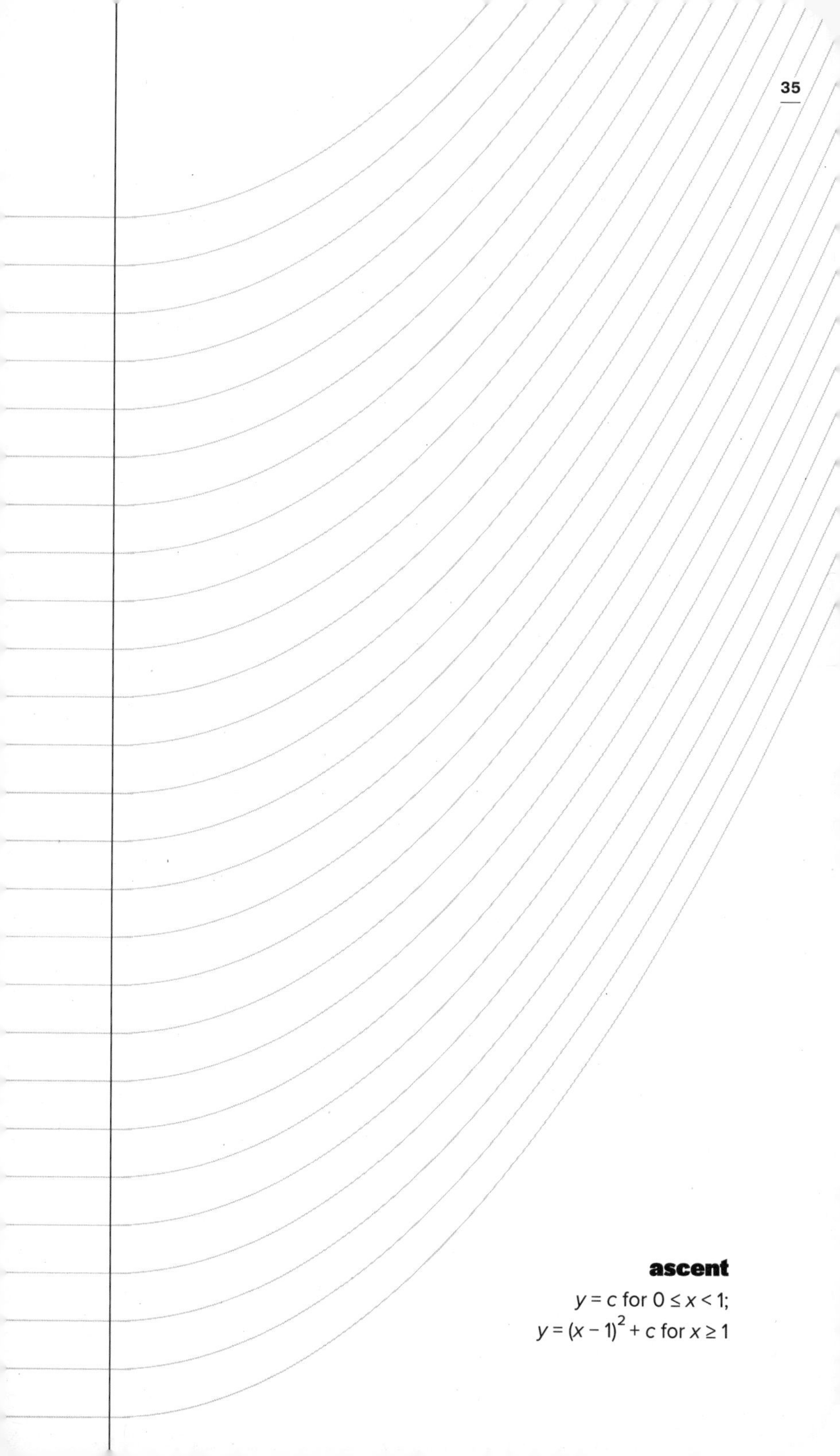

ascent

$y = c$ for $0 \leq x < 1$;
$y = (x - 1)^2 + c$ for $x \geq 1$

hill meets sky

$y = x^2 + c; y = -x^2 - c;$

origin shifted to the center of the page

rotating parabola

take the parabola in black and rotate it about a spiral

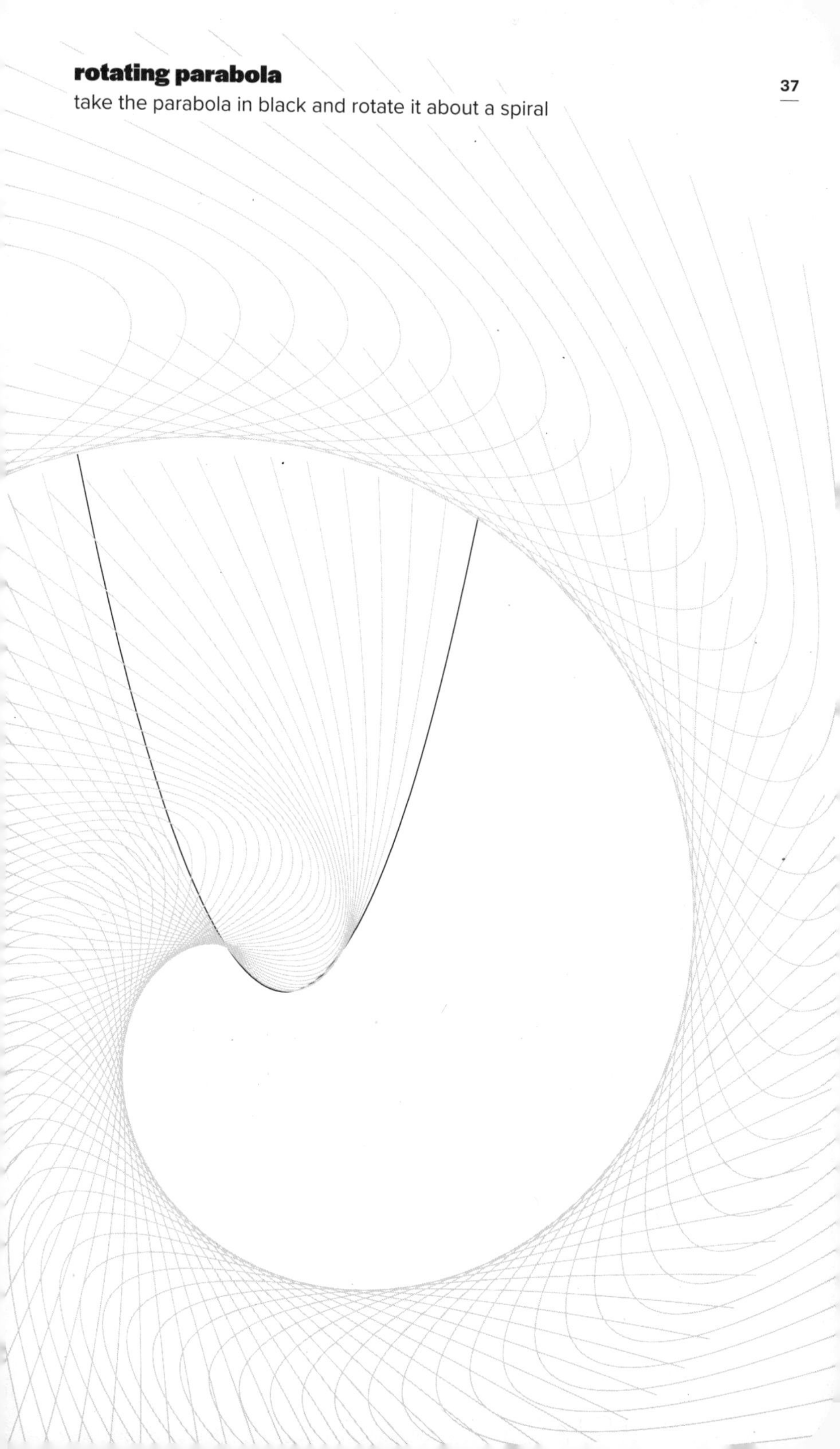

frayed

$x = (y - c)^2 - c$

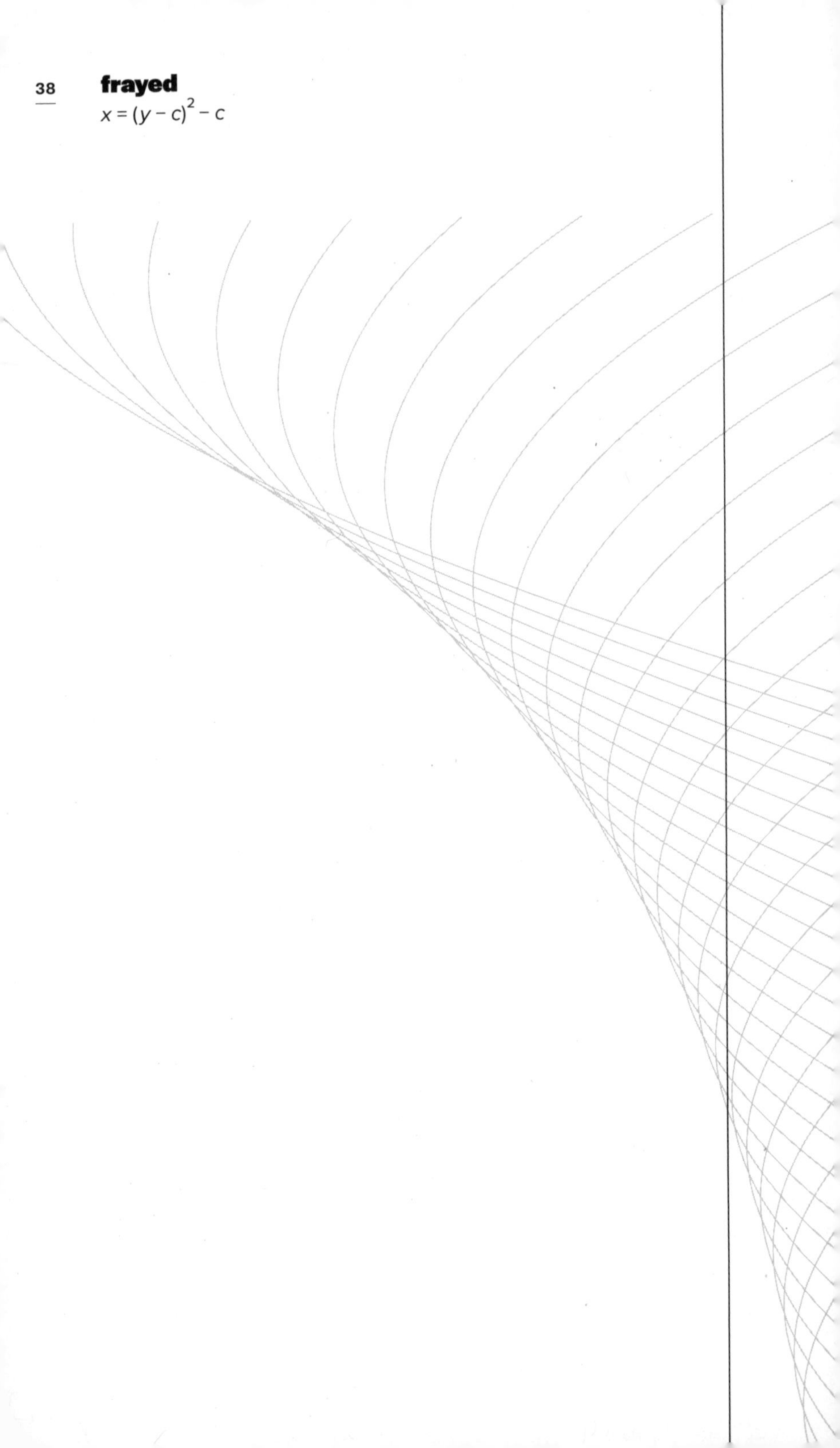

projectiles

In the absence of air resistance, a ball follows a parabolic trajectory. How might the trajectories on this page change if the balls were thrown into the wind?

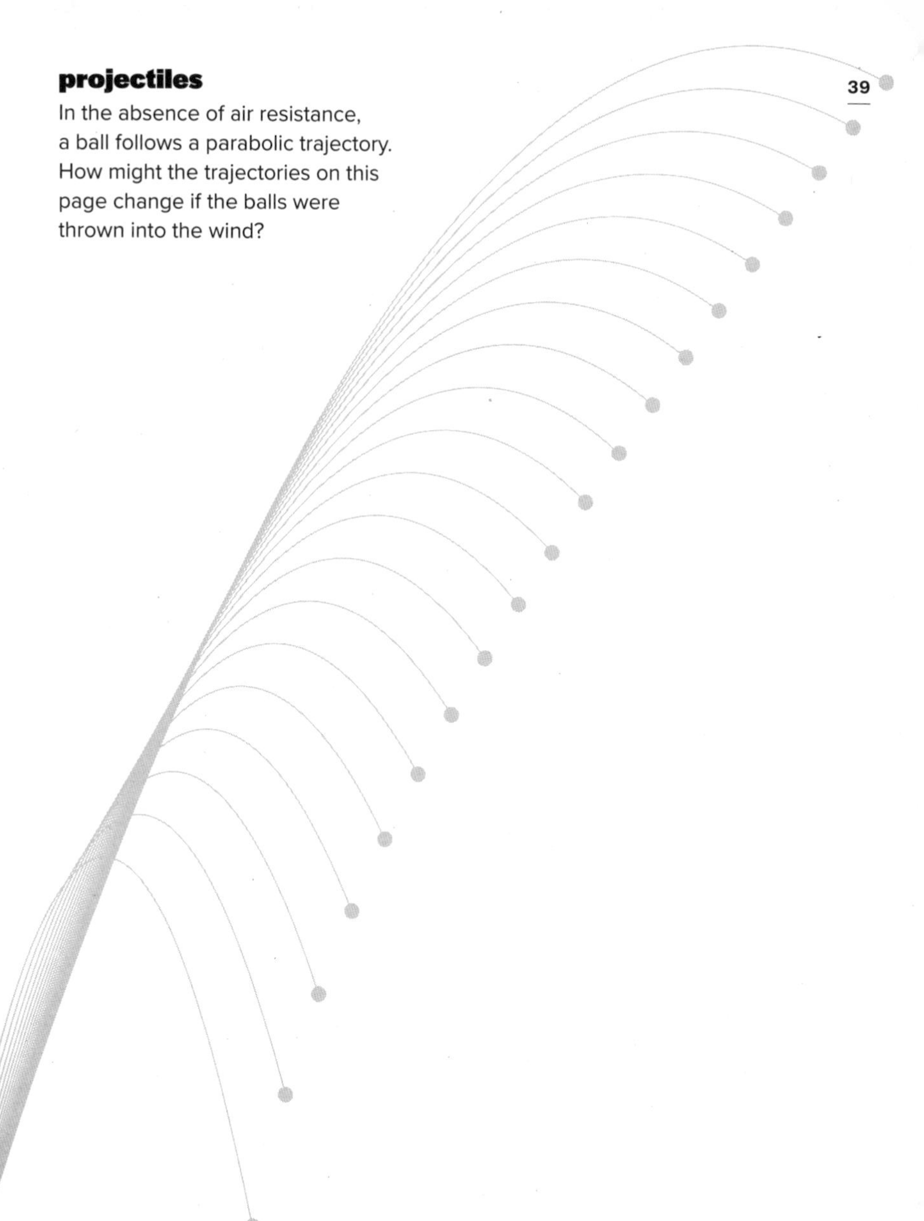

polynomial, 9 rows

mosaic with 9 rows of piecewise polynomials

polynomial, 18 rows

mosaic with 18 rows of piecewise polynomials

polynomial, 34 rows

mosaic with 34 rows of piecewise polynomials

polynomial, 70 rows

mosaic with 70 rows of piecewise polynomials

catenary

$y = (e^{x} + e^{-x})/2 + c$,

shifted to the left

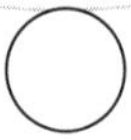

Hang a chain, a telephone wire, or a heavy cable. What shape does it form? At first glance, the curve looks like a parabola—at least, that's what Italian astronomer, physicist, and engineer Galileo observed when he tackled this problem. But glance again, Galileo. This curve is a catenary, described not by a quadratic function, in which the variable x is raised to the second power, but by a hyperbolic one, in which the number e is raised to the xth power. Not all that glitters is gold; not all that curves is parabolic.

CHAPTER: 3

Polygons the molecules of shape

It's a basic pattern of existence: small things combine to make larger things. Atoms to molecules; molecules to cells; cells to organisms. From simplicity to complexity to greater complexity, over and over again.

The same is true in math: points to lines, lines to polygons, and polygons to everything else.

A polygon is a closed figure made of straight lines. The founding polygon is the three-sided triangle, and you could say that all the later polygons are triangles at heart. The four-sided quadrilateral is two conjoined triangles, the five-sided pentagon is three conjoined triangles, and so on.

But a polygon is no mere pile of line segments, any more than you're a mere pile of atoms. Just as physics gives rise to chemistry and chemistry to biology, so too does stepping up the ladder of mathematical complexity give rise to new fields of study.

A case in point: architecture. Triangles and quadrilaterals lend their form to vast bridges, ornate opera houses, and other monuments of civilization.

Another example: computer animation. To mimic the fluidity of water or the curvature of a face, 3D animators work by increments, arranging polygons upon polygons until an effect is achieved.

And a favorite example of ours: fractals. As we'll see, these are shapes of bottomless intricacy, created by repeating a simple pattern at smaller and smaller scales, a path to infinitude paved with polygons.

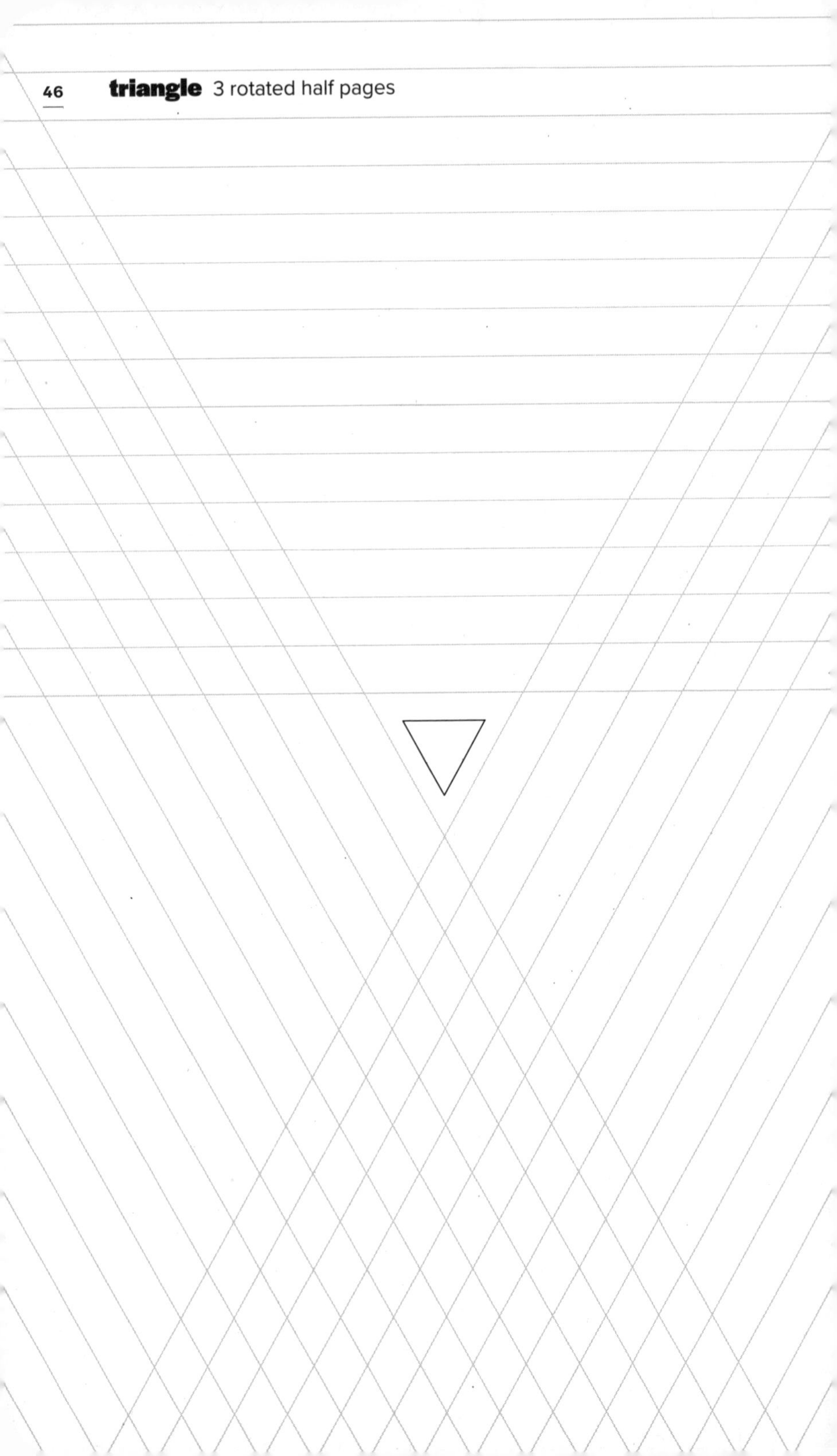

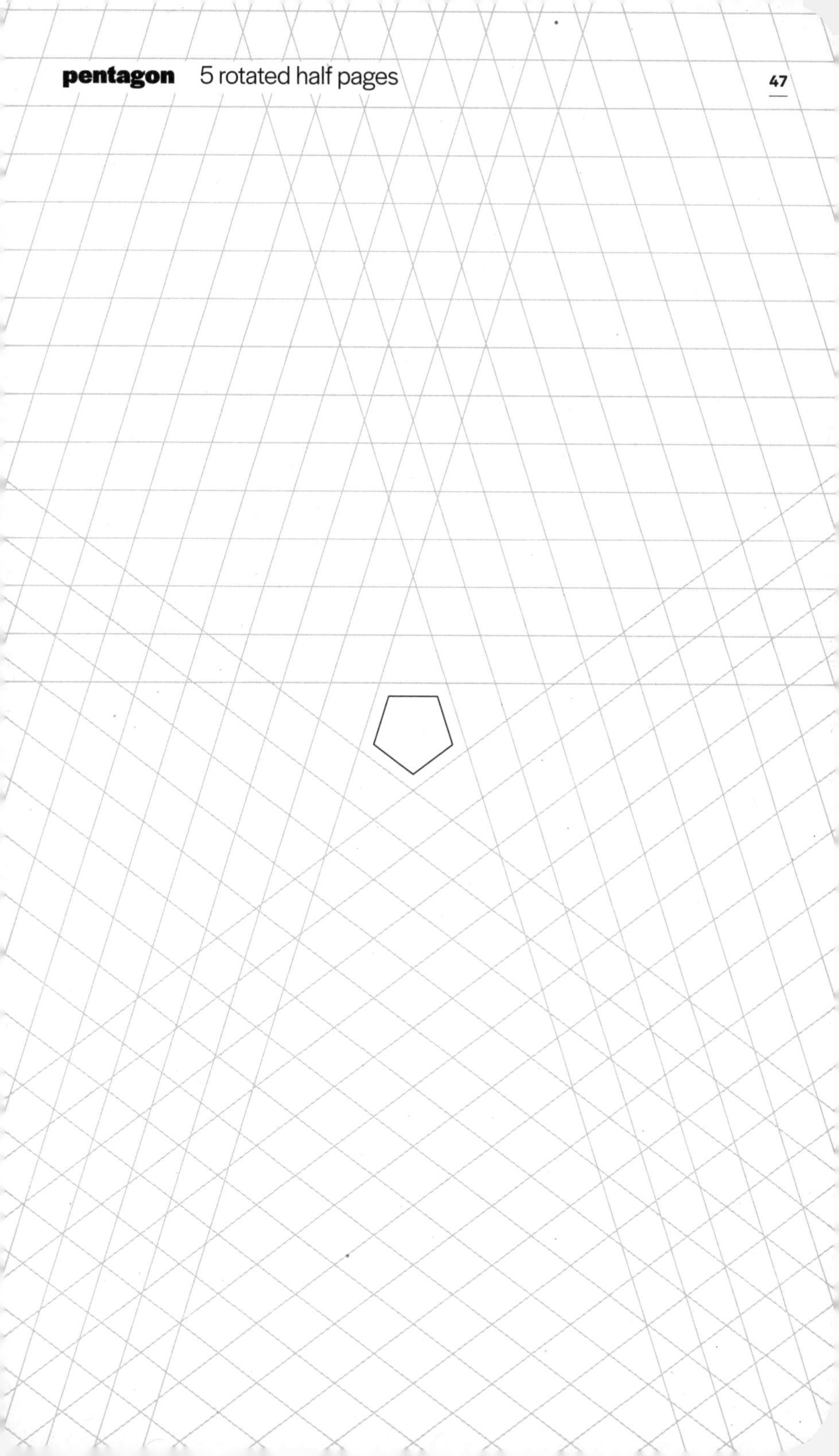

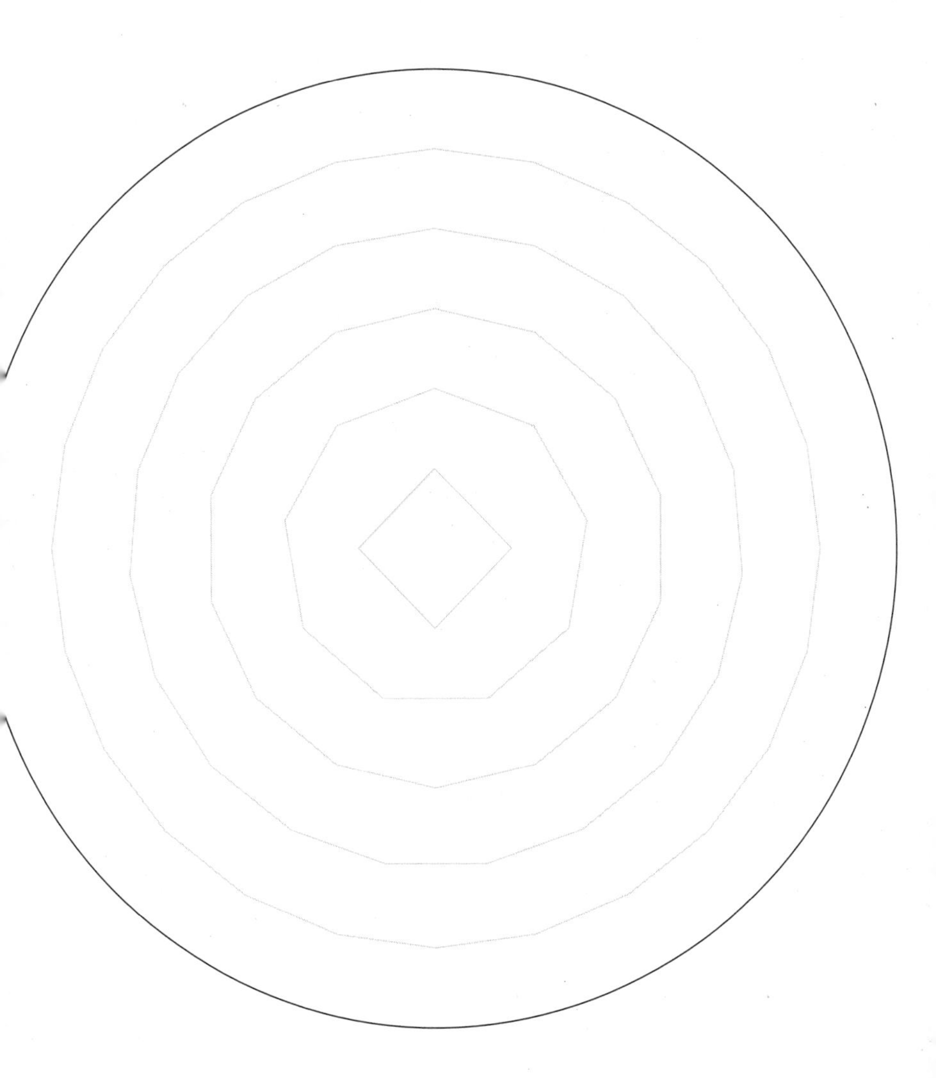

A few line segments may not look much like a curve. But the more segments you use, the smoother the appearance. On this page, the outer black line is not a true circle, but a chain of 999 line segments. The same principle underlies 3D animation: video games and movies use linear forms to approximate cars, flowers, and faces.

trio

rotate 3 triangles

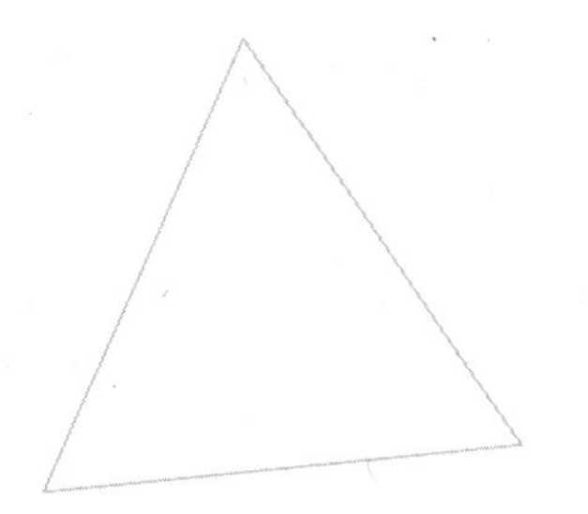

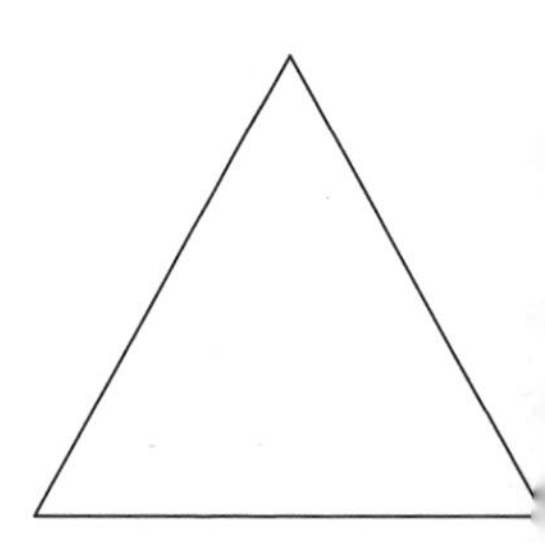

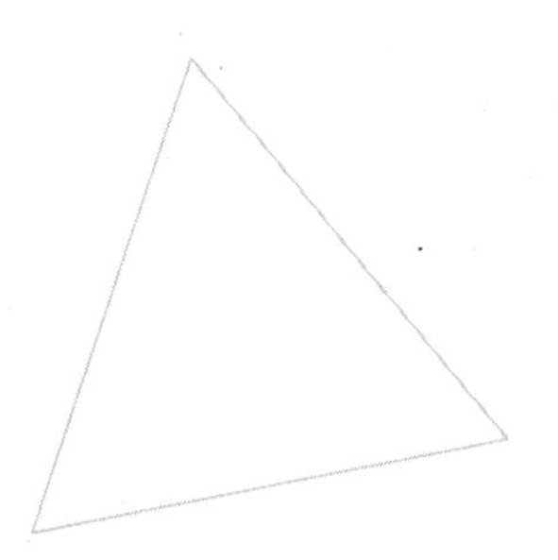

flower

rotate 10 triangles

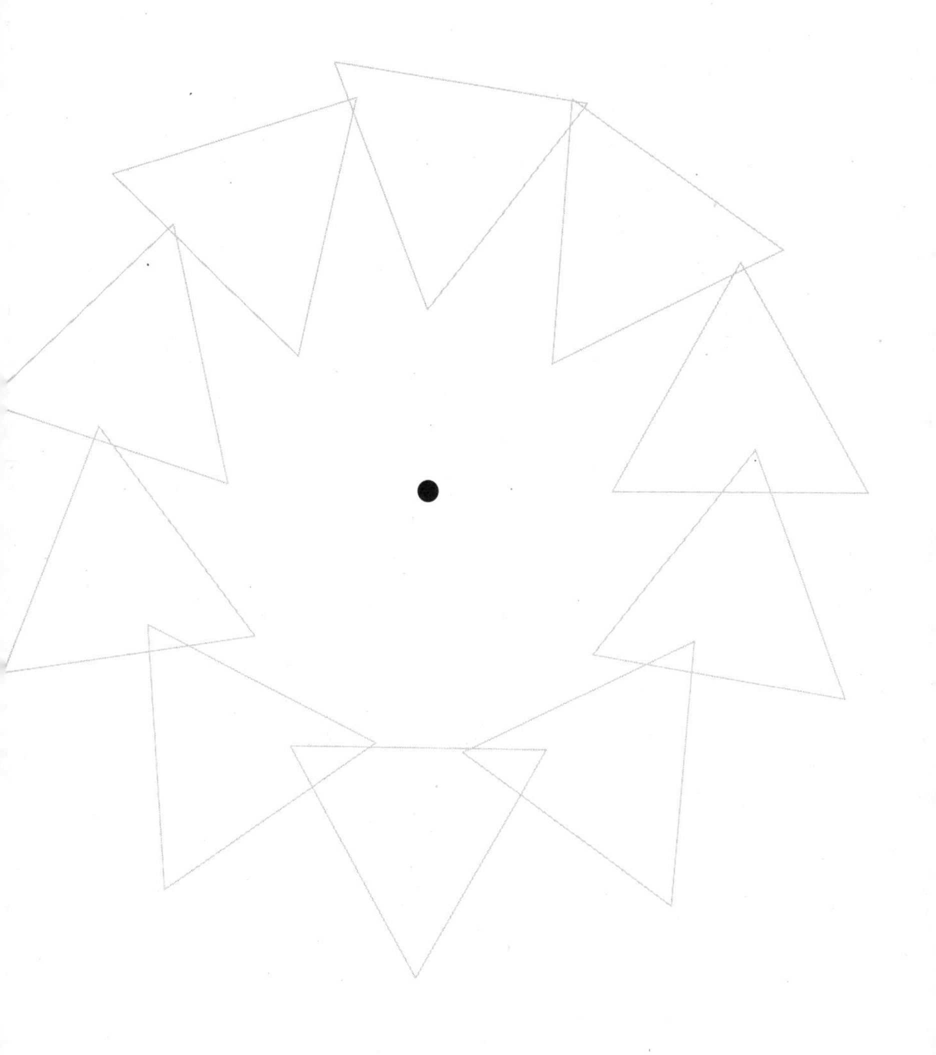

crown

rotate 30 triangles

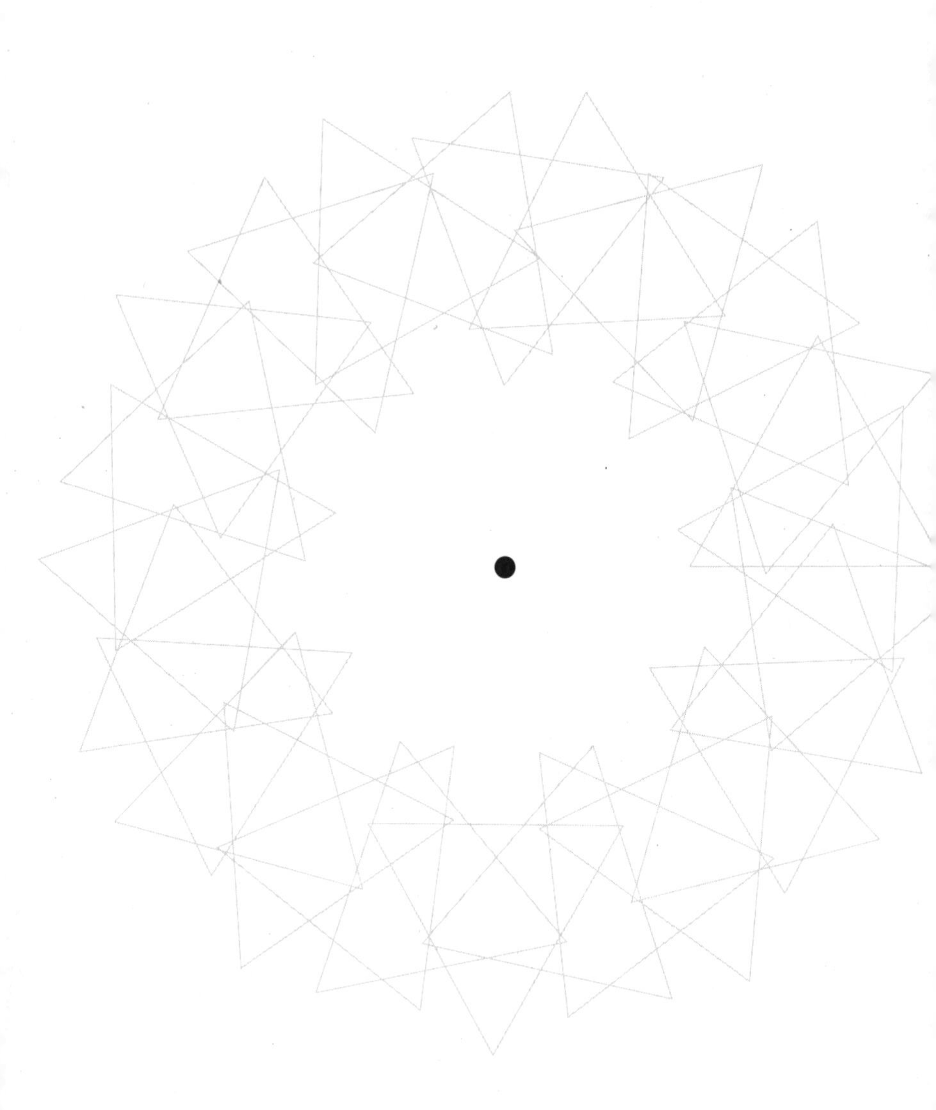

spirograph

rotate 90 triangles

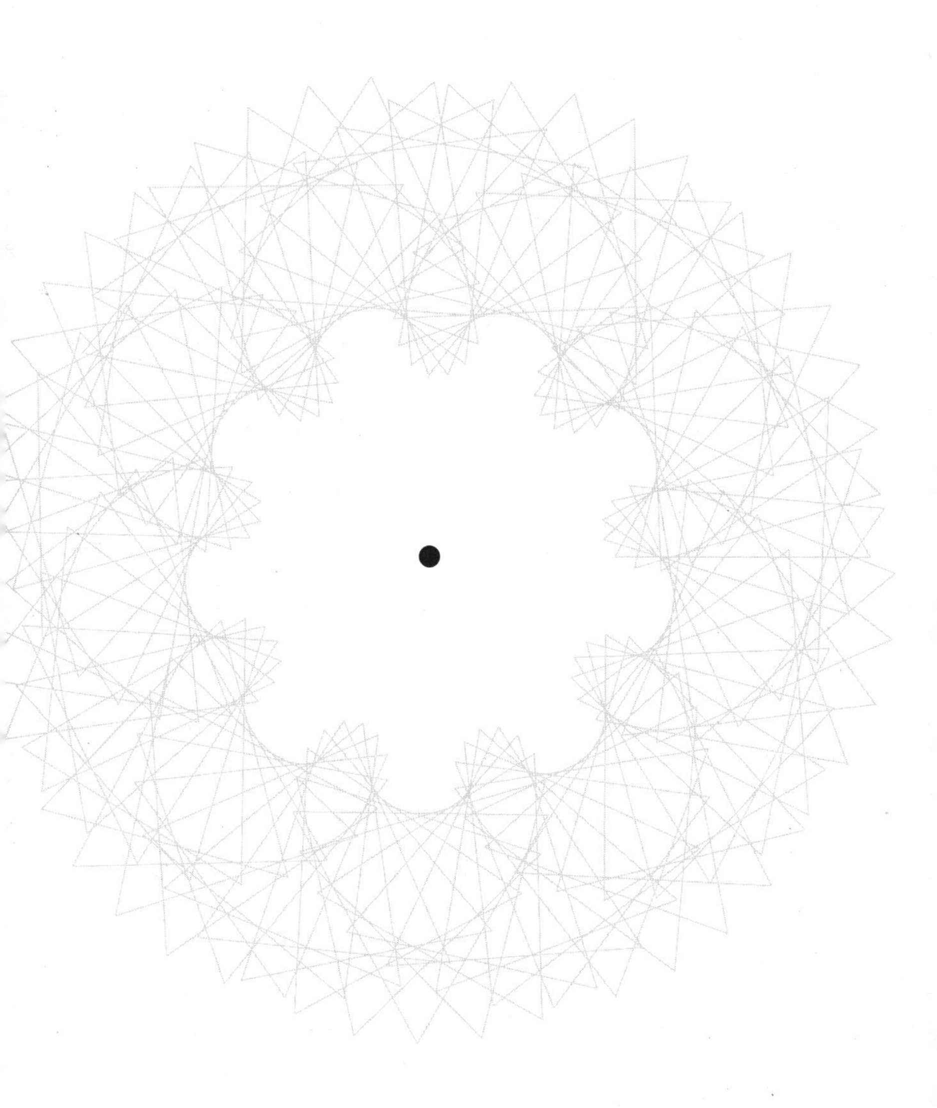

The emerging shape is called an epicycloid, which we will see again in chapter 10.

1st iteration

begin with equilateral triangles

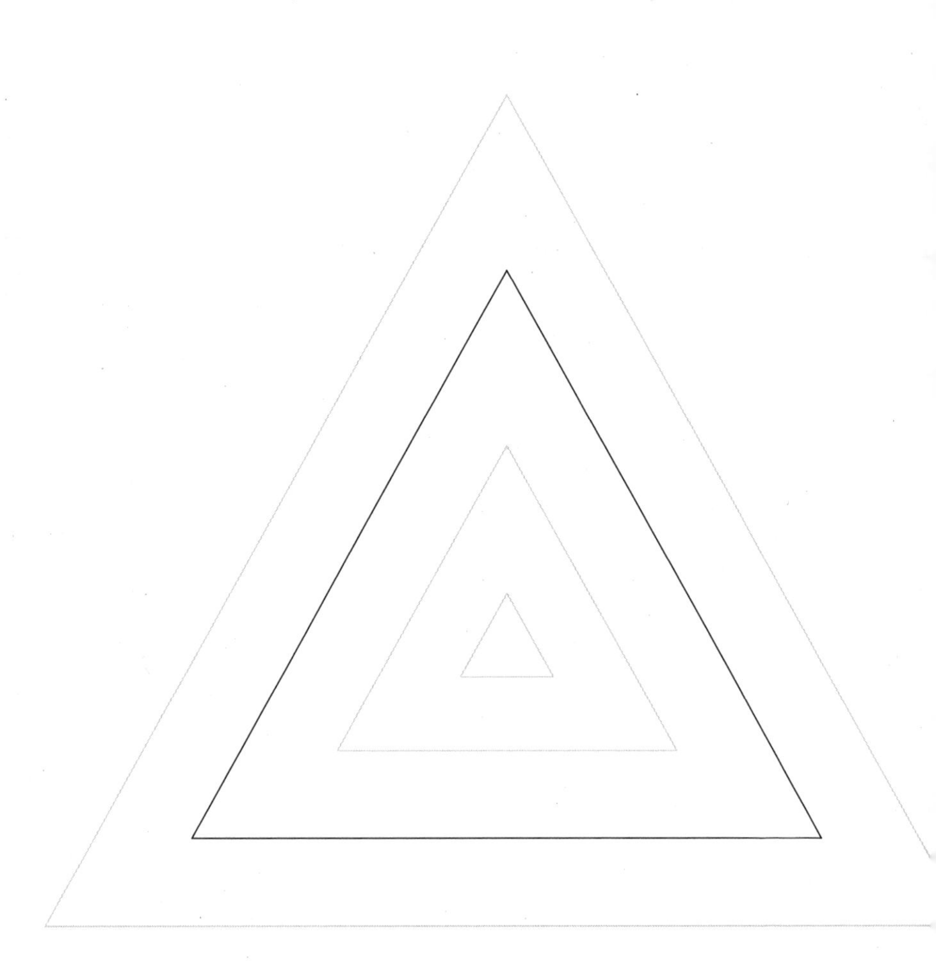

2nd iteration

on each straight line in the previous image, delete the middle third and replace it with two sides of an equilateral triangle

3rd iteration

on each straight line in the previous image, delete the middle third and replace it with two sides of an equilateral triangle

4th iteration

on each straight line in the previous image, delete the middle third and replace it with two sides of an equilateral triangle

Koch snowflake

on each straight line in the previous image, delete the middle third and replace it with two sides of an equilateral triangle

Here we see the Koch snowflake: a pattern of corners on corners on corners on corners, the same at every scale or level of zoom. This artificial creation, believe it or not, actually mimics the tiered structure of nature. Look at the splitting branches of a tree, the forking pathways of the lungs, the rough edges of a cloud, or the craggy mass of a mountain, and you will find a similar pattern: corners on corners, echoing across scales from the tiny to the vast.

CHAPTER: 4

Circles the ripples of distance

What defines a circle? None of the qualities that usually catch our attention. Not its roundness, or its symmetry, or its simplicity.

No, what defines a circle is distance.

A circle is a collection of points all the same distance from a shared center. The circle's shape is thus inextricable from a concept of distance—and the concept of distance is inextricable from the circle's shape.

The circle, you might say, is the physical manifestation of equidistance. Why does a stone make circular ripples in a pond? Because the points on the ripple are all equidistant from the site of the splash. Why do large planets and stars form spheres? Because that way, the points on its surface are all equidistant from the gravitational center. "Life is a full circle," wrote author Anaïs Nin, "widening until it joins the circle motions of the infinite."

3 circles

a circle, repeatedly rotated 120° about the black point

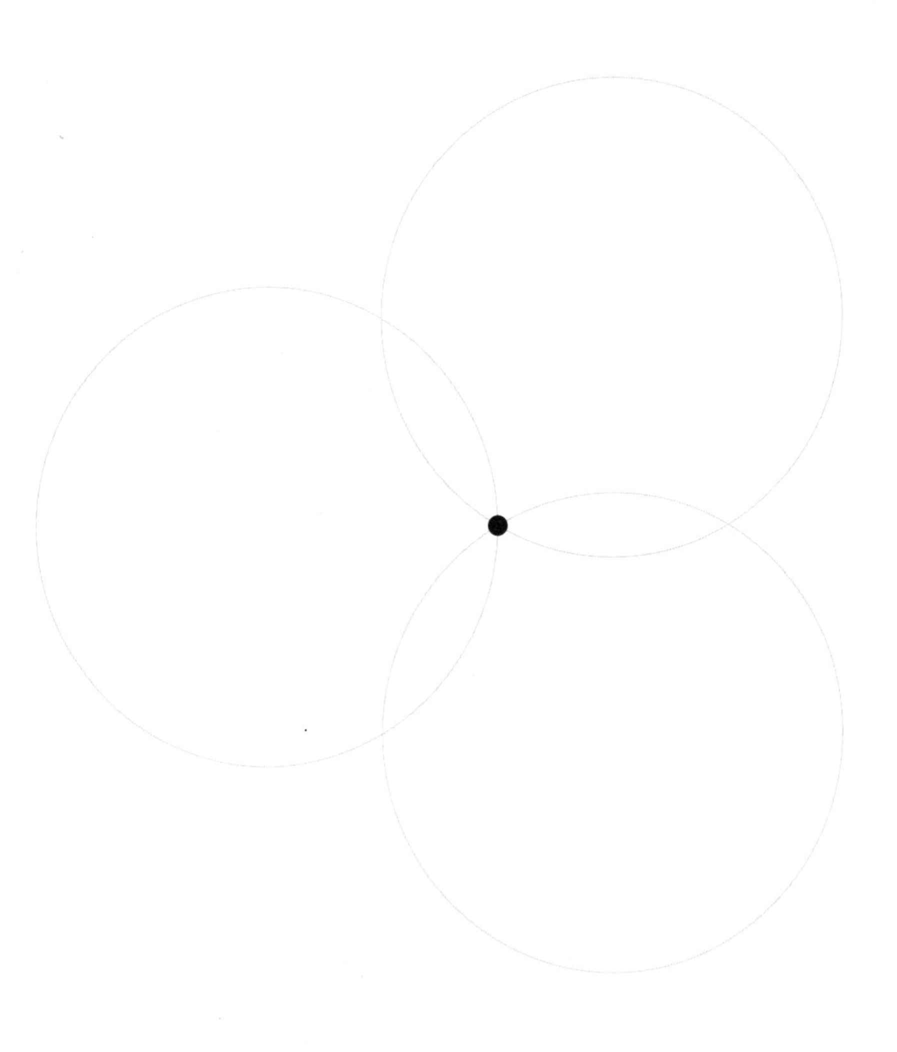

6 circles

a circle, repeatedly rotated 60° about the black point

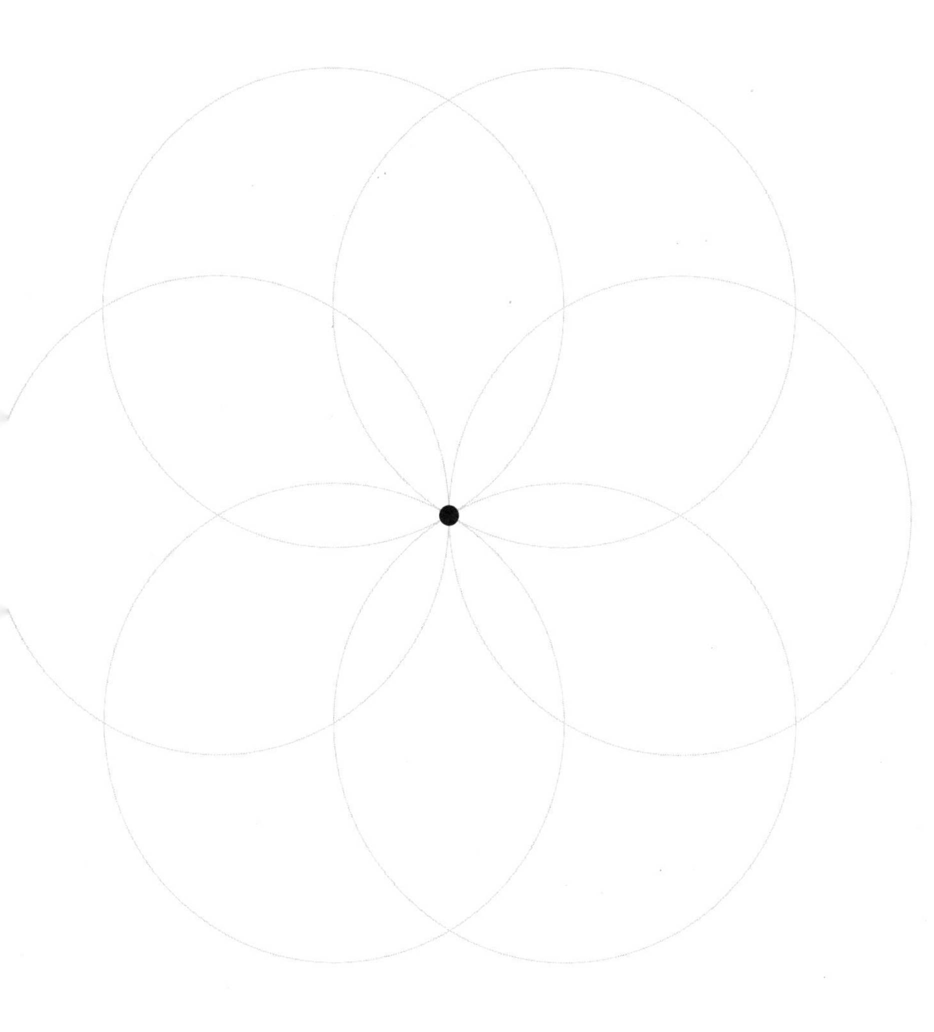

12 circles

a circle, repeatedly rotated 30° about the black point

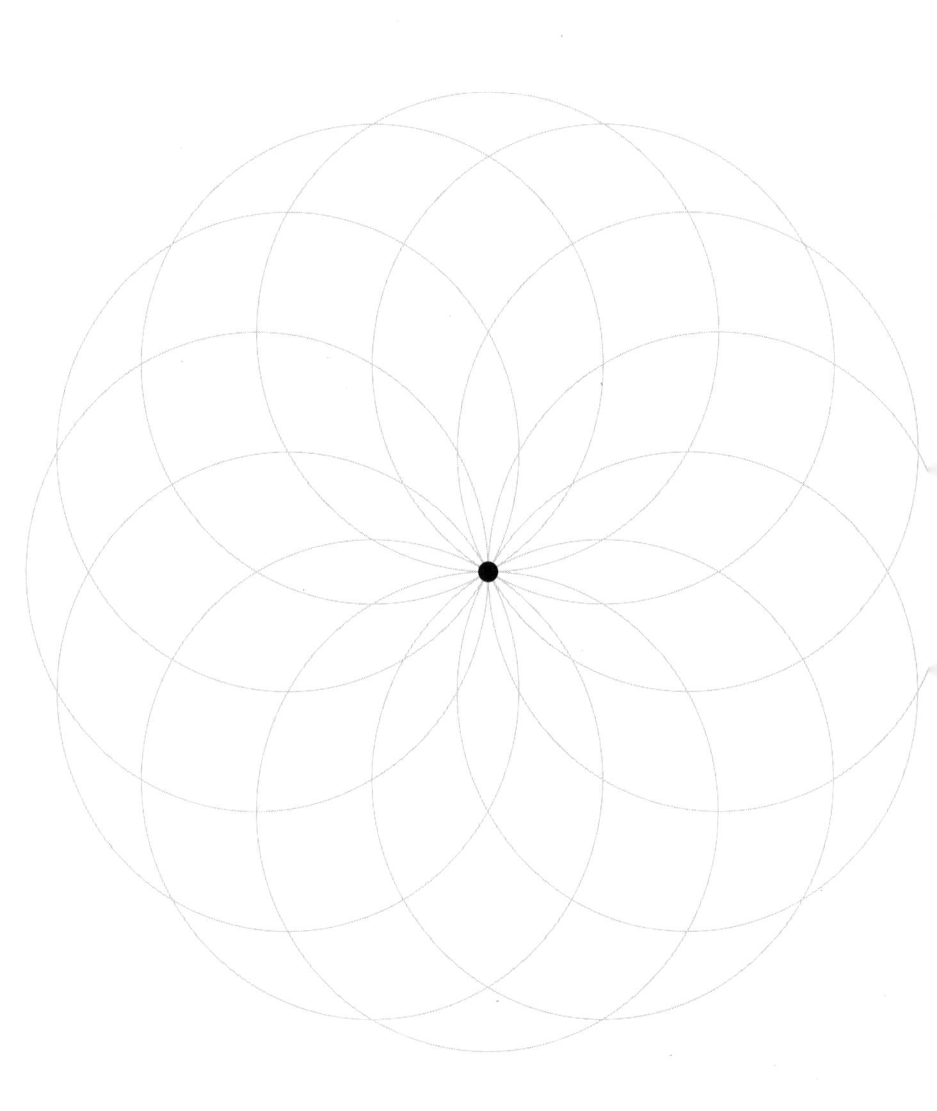

32 circles

a circle, repeatedly rotated 11.25° about the black point

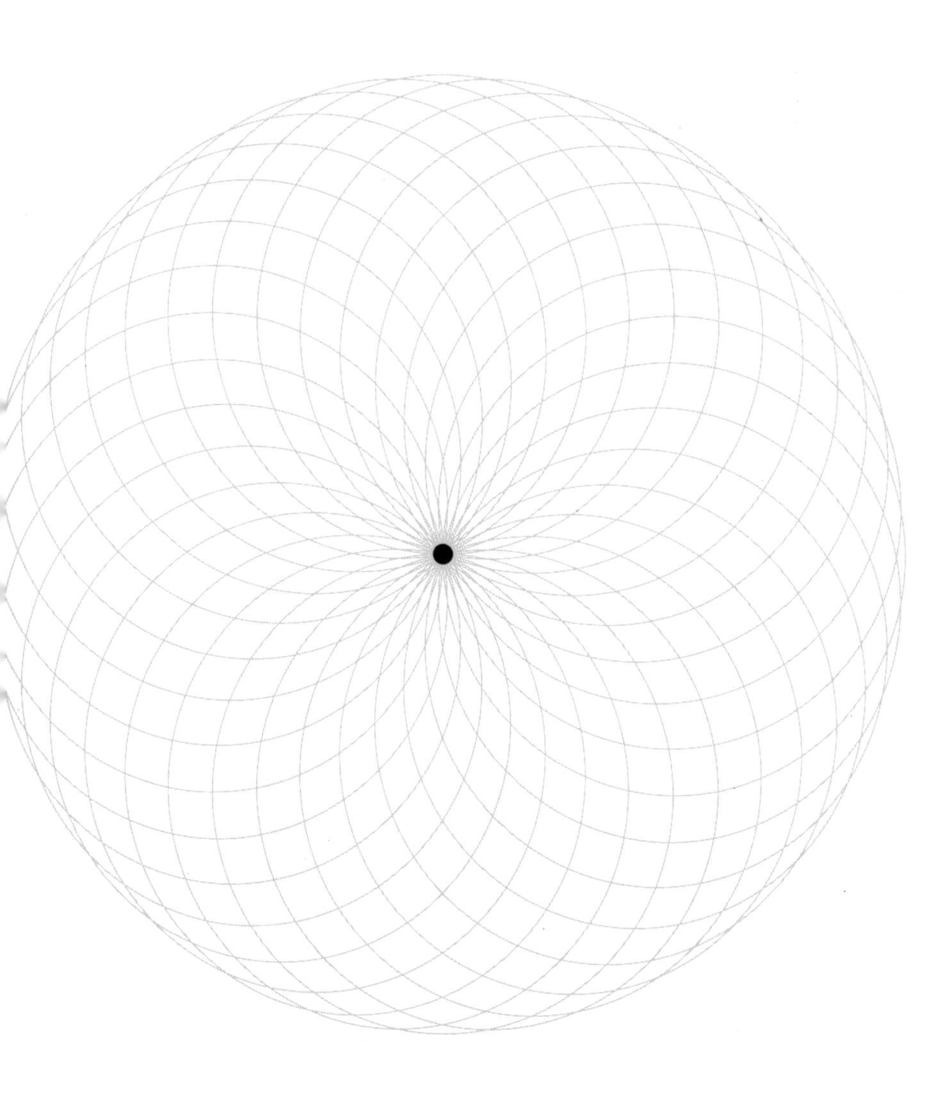

ripples

$y = \sqrt{c^2 - x^2}$ for $x \leq 0$

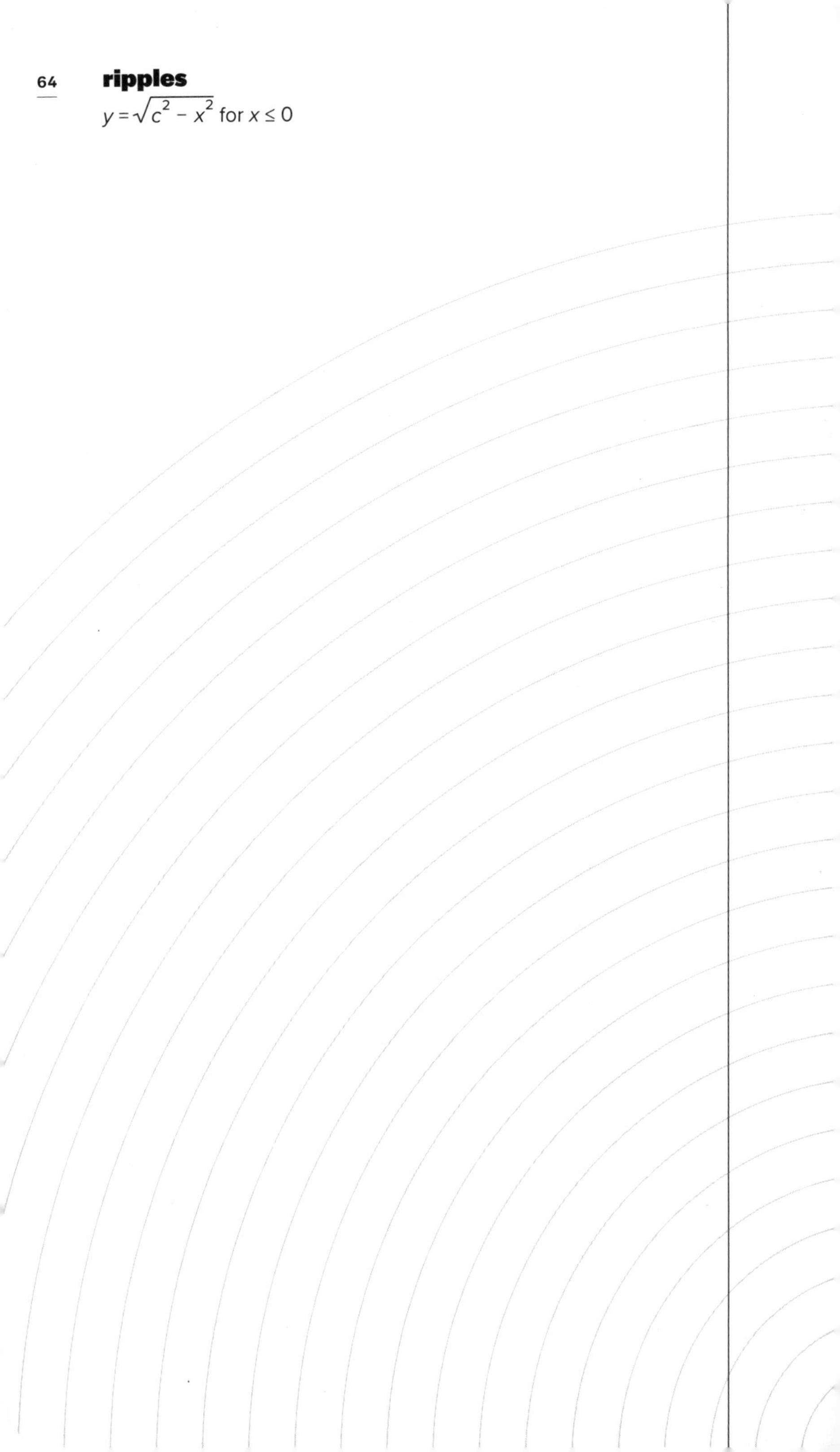

the yo-yo

$y = c$ for $0 \le x \le 0.194k$; $(x - 0.194k)^2 + (y - c + 0.14)^2 = (0.14)^2$
for $0 \le k \le$ number of circles

two submarines

circles centered at bottom right and top left, with radii differing by the standard line height (0.28 inches)

cloud cover

$y = c$; draw white circles with radii equaling 1.75, 1.75, and 2.5

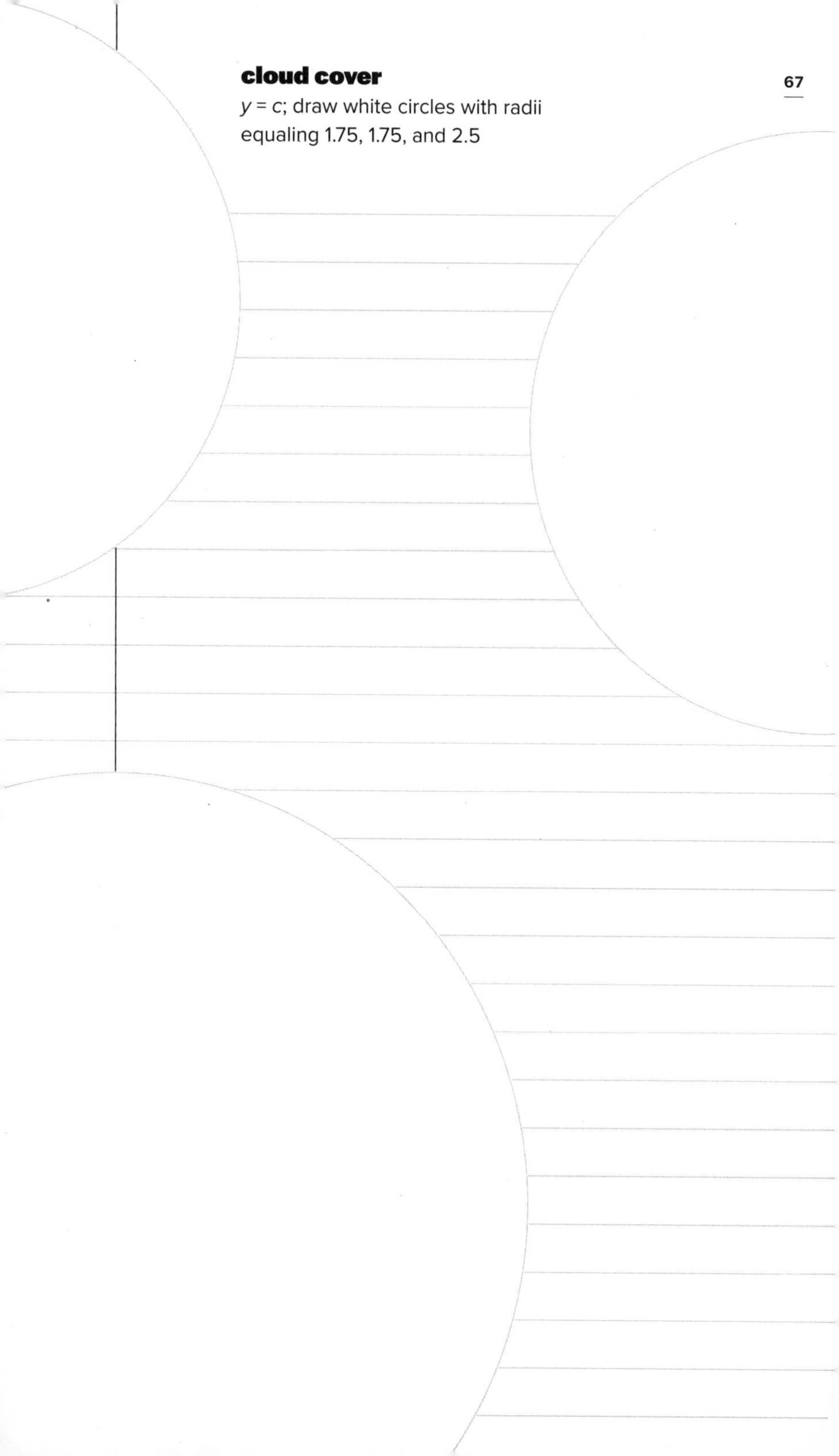

three-hole punch

create a standard page; copy a large circular region from the center of the page; rotate 90°; place three copies

bird on a wire

to create the kth line, $y = c$ except when $k = 13$; when $k = 13$, connect the endpoints of $y = c$ with a lower point slightly off-center horizontally; draw a white circle over the bent line

waterfall

2 horizontal lines connected by a quarter circle with a radius of 3 line heights

nowhere square

draw 4 sets of concentric circles;
overwrite with a white square

auditorium

circles centered horizontally on the page and 1 line height above and below the center of the page vertically

seashell

begin with the circle in black; for each step,
rotate and enlarge the circle at the previous step

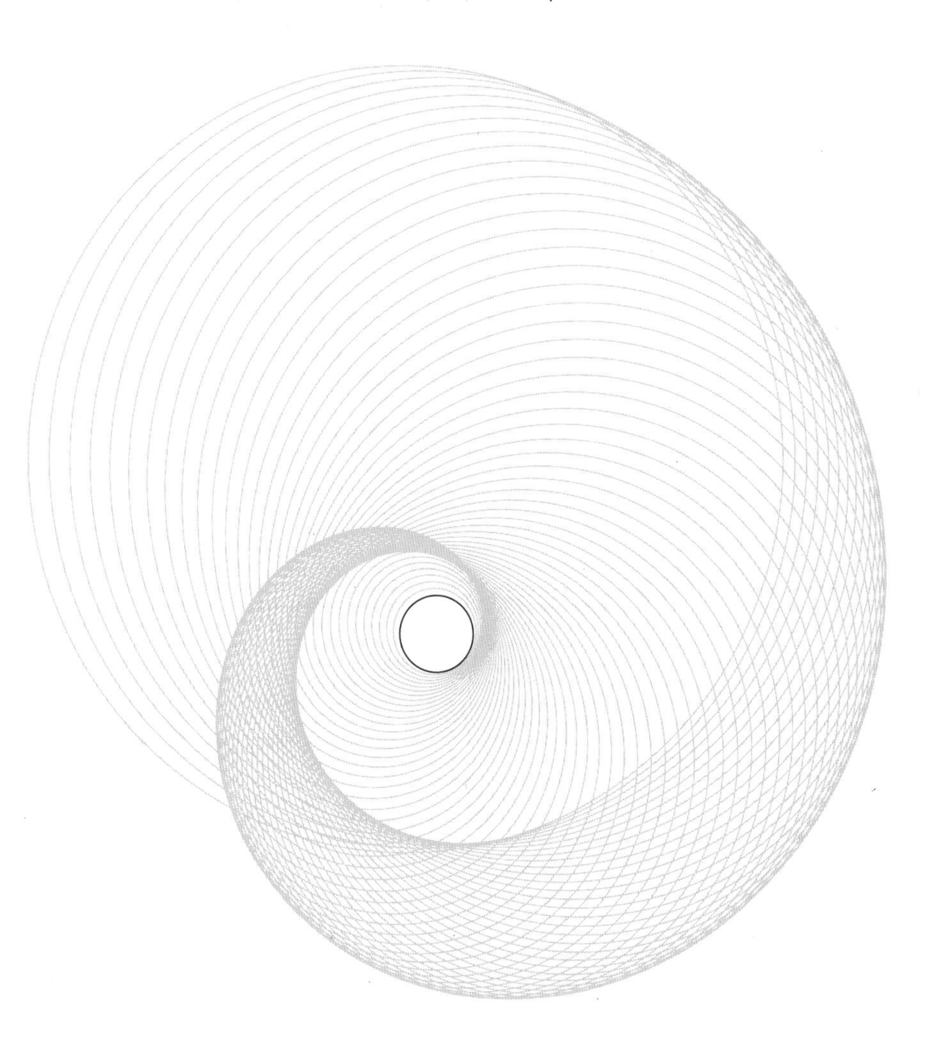

speedbump

semicircle with radius = $1.1k + 0.3$, centered at $(-1, c)$;
when the semicircle intersects the kth line, continue the line

tardis begin with the circle in black; for each step, rotate and enlarge the circle at the previous step

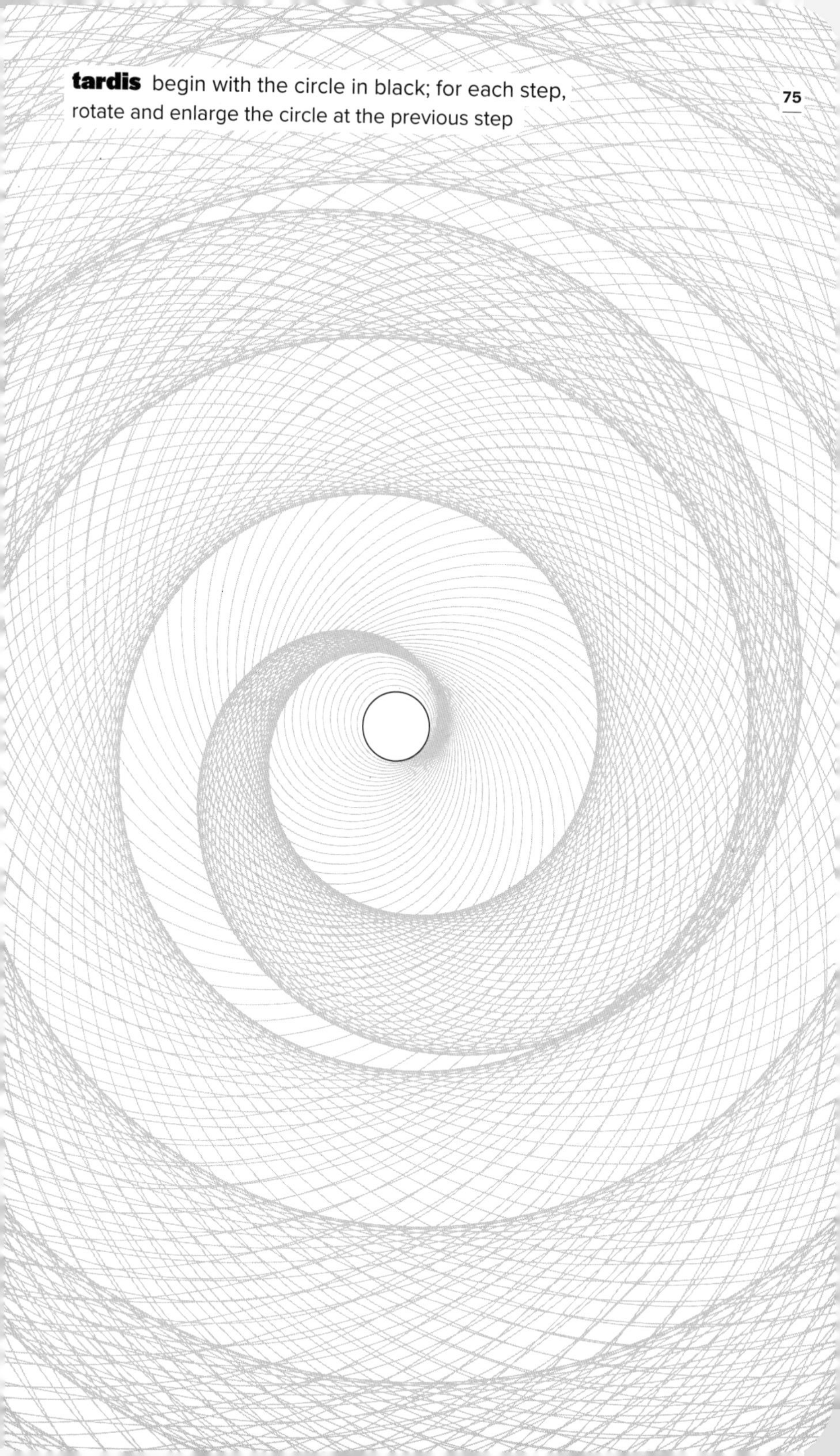

Reuleaux

draw an equilateral triangle; round the edges by drawing circular arcs centered at one vertex of the triangle and passing through the other two vertices

This shape takes its name from 19th-century German engineer Franz Reuleaux, who quite literally reinvented the wheel. How so? Well, most shapes, such as triangles and squares, would make bad bicycle wheels—not just because they don't rotate well, but because you'd be bumped up and down as they rotated. Not so with a circle: it gives a smooth ride because, as it rotates, the diameter never changes. Surprisingly, the same is true of the Reuleaux triangle, whose width is constant. Thus, a bicycle with Reuleaux wheels would ride as smoothly as an ordinary bicycle.

CHAPTER: 5

Waves the rhythms of nature

"Waves are the practice of water," wrote the Zen monk Suzuki Roshi. "To speak of waves apart from water, or water apart from waves, is a delusion." Perhaps so. But to a mathematician, the practice of waves goes far beyond water. The wave is the primordial form of repetition. It is cyclicity incarnate.

The wave is best understood as a circle unraveled. Take a spinning Ferris wheel, fix your eyes on a single point, and track its height as it rises and falls, rises and falls.

Up, down . . .

Up, down . . .

Up, down . . .

These are the ups and downs of the sine curve, the metronome of the natural world. From sunrise to sunset, from summer to winter, our life moves in cycles and waves. Music is waves of sound; color is waves of electromagnetic radiation; atoms are waves of quantum potentialities. Even language and memory can mimic the quality of waves. As Virginia Woolf put it in her novel *The Waves:* "Words that have lain dormant now lift, now toss their crests, and fall and rise, and fall again . . ."

cups and bowls

alternate upper and lower semicircles with centers along $y = c$

synthesizer

x points evenly spaced along the spine; y points alternating between c and $c + 0.3$; for each c, connect (x, y) to form lines

choppy seas

$y = \sin(x) + c$

polygraph

four copies of $y = e^{-(x-2)^2} \cos\left[30(x-2) - 6.5(x-2)^2\right]$, shifted up and to the right

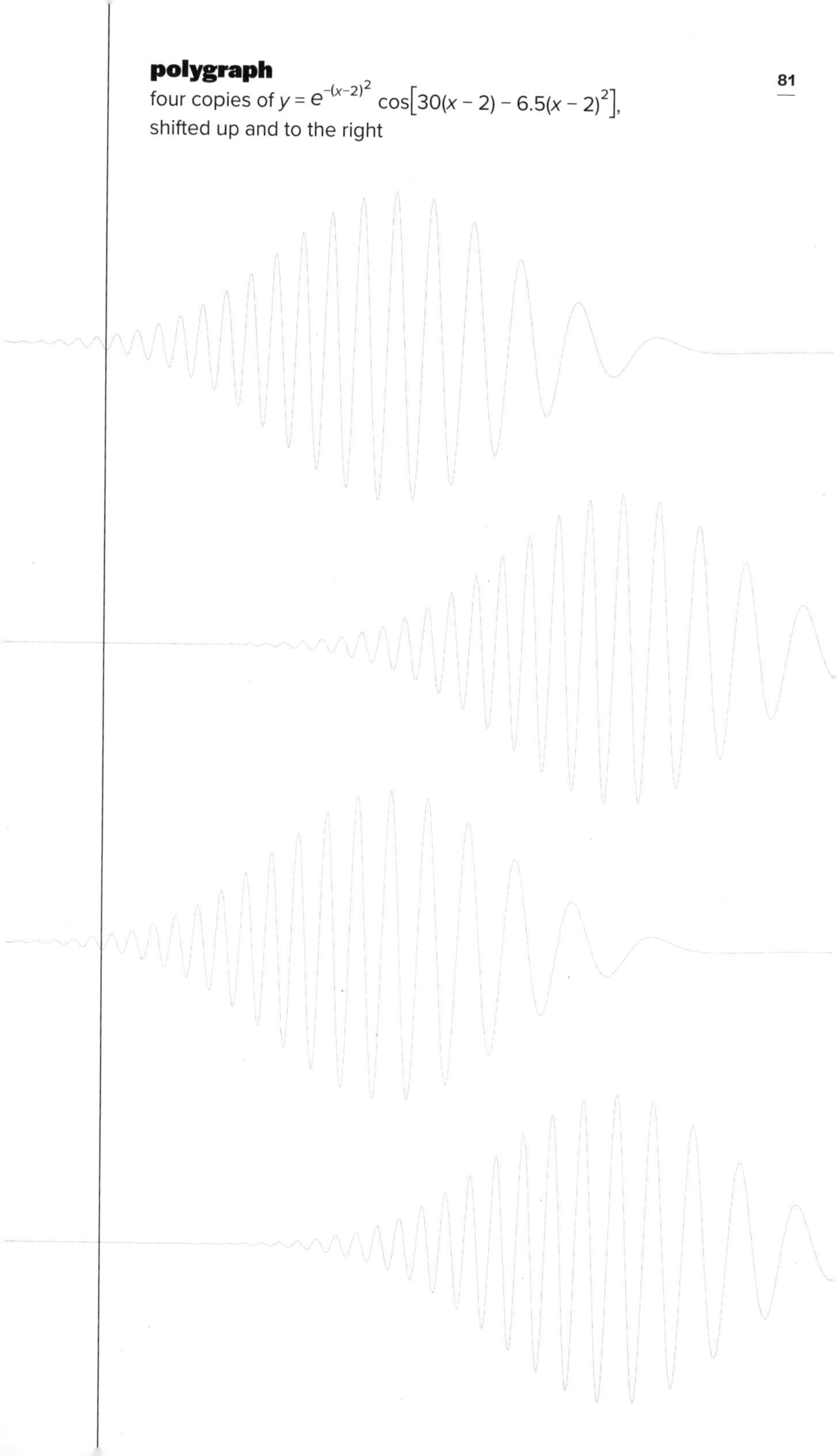

braids

$y = \sin(x) + c$; $y = \sin(x - \pi/3) + c$; $y = \sin(x - 2\pi/3) + c$

countryside

$y = 0.4 \sin(0.2cx) + 0.3c$

geology

$y = \sin(2x)\ \ln(-0.5x) + c$

rainstick

$y = 0.18cx + \sin(2x)$, shifted up and to the right

marshmallows

$\frac{\cos(x+y)}{20} - \cos(x) - \cos(y) = 0$, scaled down

ocean

$y = \sin(x) + c; y = \cos(x) + c$

electrified hair

$y = \sin(2x)e^{0.25x} + c$; models the height of a bouncing spring

river border

alternate between $y = c/2$ for the full width of the page and $y = c/2$ for $0 \leq x \leq \sin(c)/2 + 1$

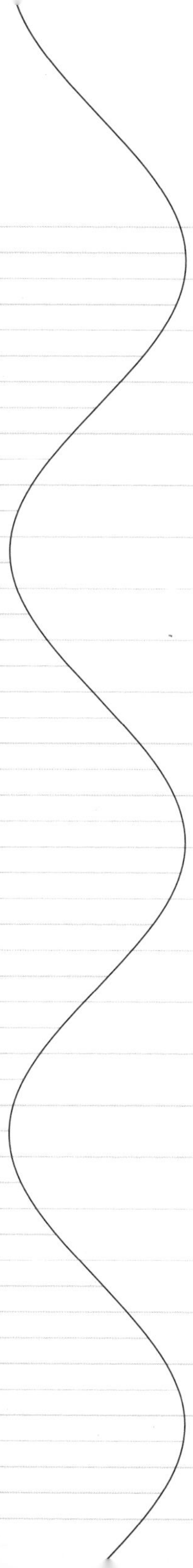

music

$y = \sin(2kx) + c$, where k is a note in Hertz; middle C on a piano is 261 Hz

Sound waves vibrate the air. The taller the wave, the larger the vibration—and the louder the note. The narrower the wave, the faster the vibration—and the higher the note. For example, the note middle C corresponds to $y = \sin(522\pi x)$, and the slightly higher note D can be described as $y = \sin(660\pi x)$. This page is a kind of mathematical sheet music: each line corresponds to a note (all played for the same duration and at the same volume) in “Mary Had a Little Lamb.”

CHAPTER: 6

Limits tales of striving

We tend to imagine infinity as a distant, gargantuan thing. But to a mathematician, infinity is more pervasive than that—and more unsettling.

"A long stick," begins a riddle that circulated in ancient China. "Each day, take away half. In a myriad ages, it will not be exhausted." But how can this be? Shouldn't the stick eventually vanish? What does it even mean to have a stick so small that we can't see it?

You may have heard the same idea presented as Zeno's paradox. To cross a room, you must first go halfway. After that, you must cross half of the remaining distance. Then you must cross half of the distance that still remains. On you must go, in an endless sequence of halfway crossings. How, then, can you ever arrive?

These riddles long vexed mathematicians. But today, vexation has become jubilation. The idea of infinite approach, of striving and striving and striving but never reaching—this is the engine that powers modern mathematics. In the coming pages we will see lines growing ever closer, spirals growing ever tighter, waves growing ever faster. The key idea is the limit: a destination that we approach but may never reach.

The limit is the Chinese stick, entirely gone. It is Zeno's room, finally crossed. It is the fantasies and riddles of ancient mathematics, brought to life in modern minds.

limit

$y = c$, where $c = 0.28$ for the first line, and c is placed half the distance between the previous line and $y = 7.25$

pulse

$y^2 = x^2 \sin^2(15x)$, shifted up and to the right

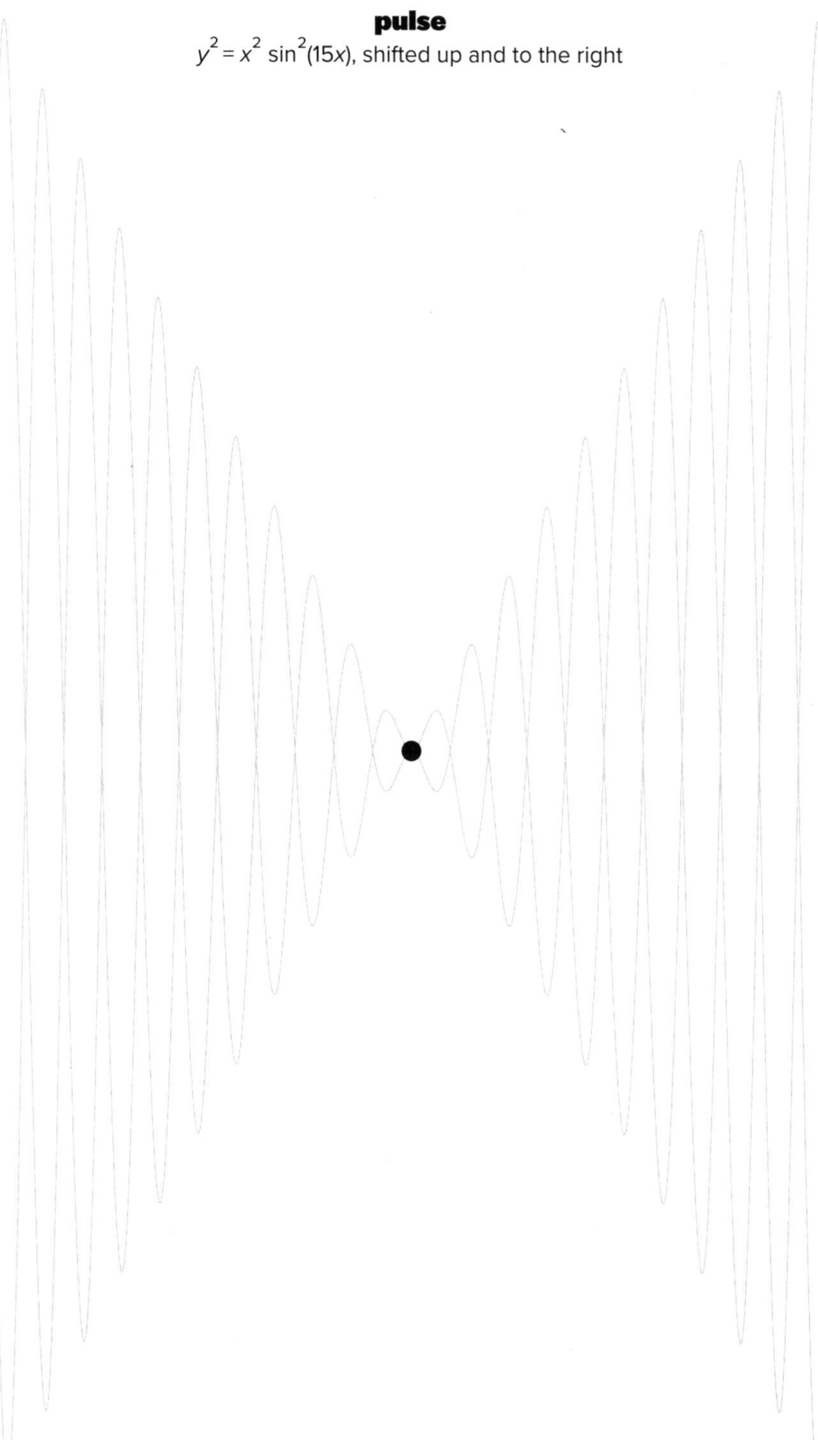

singularity
$y = -\frac{1}{x+1} + c$

coiled

black curves based on $x = \sin(1/y)$; horizontal lines are drawn between the black curves

spines
$y = \ln(-x - 1) + c$ for $x < -1$;
$y = \ln(x + 1) + c$ for $x > -1$

rupture

$y = \frac{1}{1-x} + c$ for $x < 1$; $y = \sin(x)\ \frac{1-x}{(x-1)^2} + c$ for $x > 1$;
both with asymptotes at $x = 1$

veering

$$y = \frac{x^2 - 4}{x + 2.75} + c$$

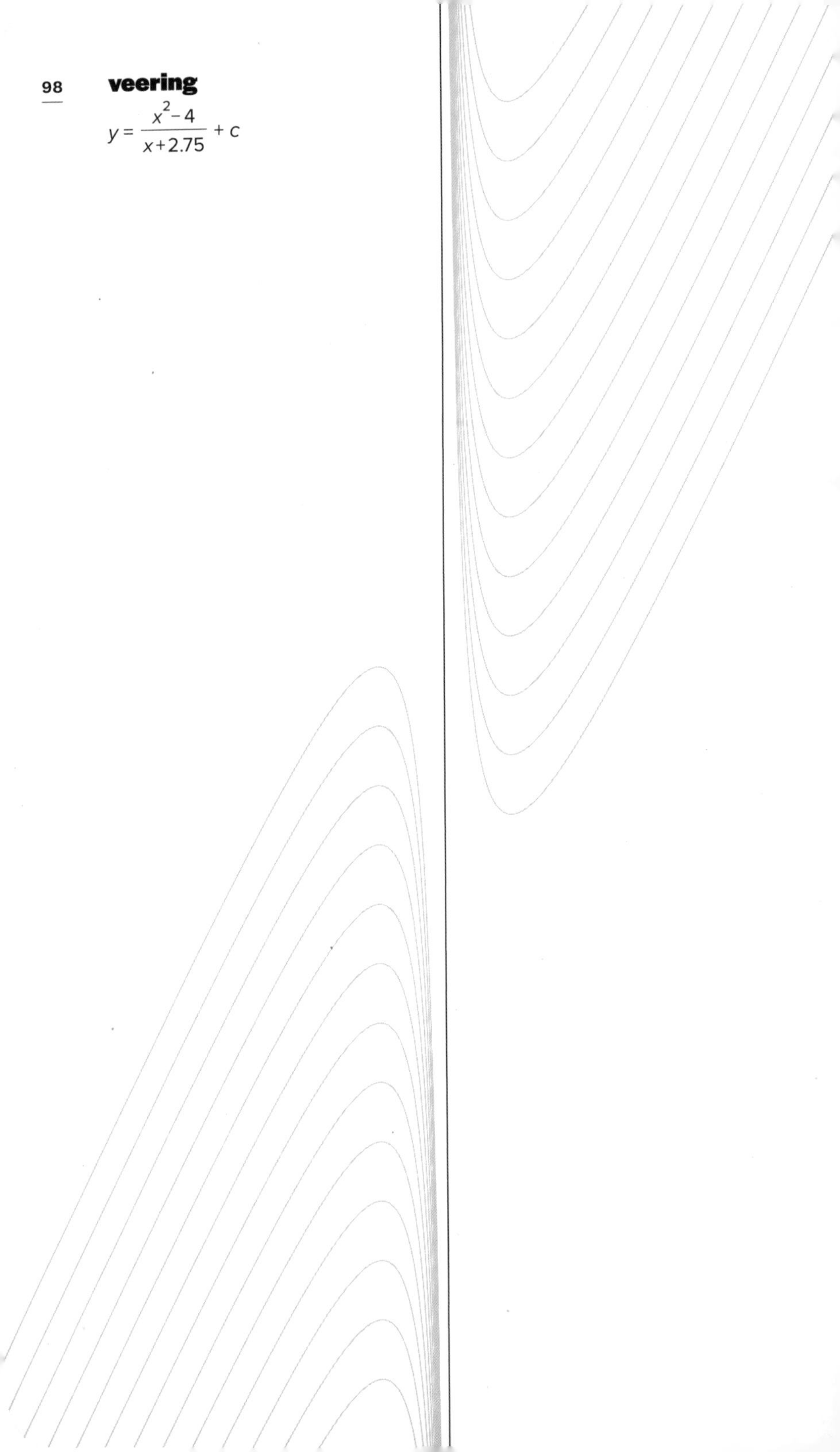

blinds drawstring is $x = \sin(1/y)$, shifted up and to the right; lower lines connect the points $(0, c)$ and $(5.5, c)$ with the relative minimum and maximum of the shifted $x = \sin(1/y)$; remaining lines are the standard $y = c$

infinite muscle

$y = -\tan(x) + c$

$$y = \frac{x+c}{3x+2} \text{ for } c \geq 2.5$$

Will the bottommost gray curve,

$y = \frac{x+2.5}{3x+2} + \frac{11}{12}$, ever touch the black line, $y = \frac{5}{4}$?

whirlpools

$r = 0.2/\theta$, shifted up and to the left repeatedly; the large whirlpool is rotated and shifted

growth

$$y = \frac{1}{1+e^{-4x}} + c, \text{ shifted to the right}$$

The logistic function, as in this example, has many applications. For instance, under chess's Elo rating system, it's used to model the chance one chess player has of beating an opponent. Because the curve approaches a height of 1 at the right and 0 at the left but never quite reaches either, this suggests that you may have *almost* a 0% or 100% chance of beating your opponent, but the game's outcome is never certain.

wings $x = -e^{cy/12}$ and $x = -e^{-cy/12}$ for $c = 1, 2, \ldots, 26$, shifted up

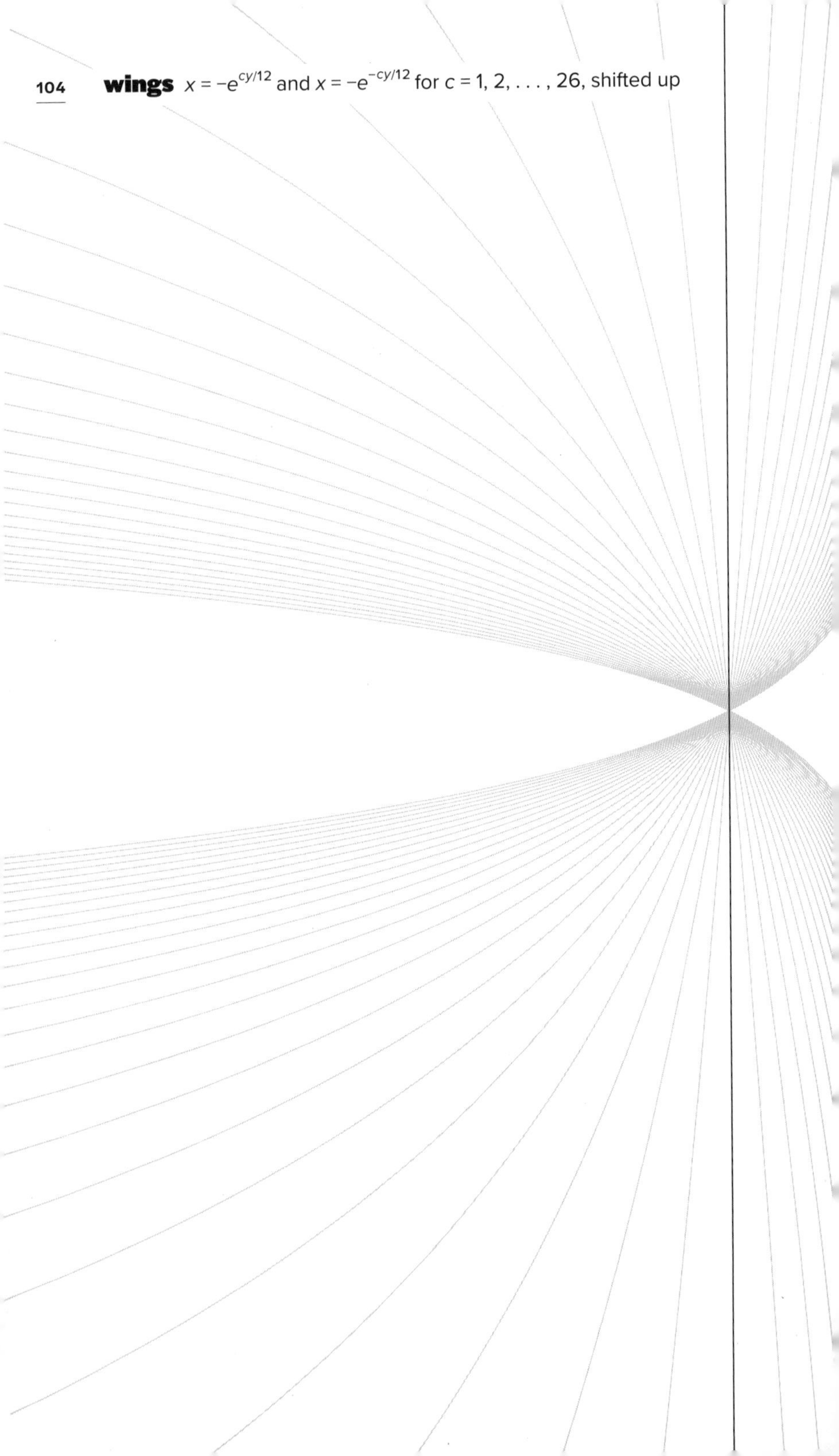

jump $y = x + c$ for $x < b$; $y = x + c - 0.125$ for $x \geq b$

The variable b is a random point between 0.5 and 4.5. When $y = x + c$ for $x < b$ and $y = x + c - 0.125$ for $x \geq b$, the graph has a jump at $x = b$, which is called a jump discontinuity.

false wave $y = x - \frac{x^3}{6} + \frac{x^5}{120} - \frac{x^7}{5{,}040} + c$, shifted to the left; polynomial approximation of $y = \sin(x) + c$

Is this a wave? No, not really. A wave rises and falls endlessly; these curves rise once and fall twice. Still, just as a many-sided polygon is the beginning of a circle, this wiggle is the beginning of a true wave. Keep adding terms to the equation, and the curves will develop more hills and valleys. Not a true sine wave, but closer and closer to one, until, in the limit . . .

CHAPTER: 7

Rotation dizzying symmetry

Have you ever ridden a spinning carnival ride? It thrills children; it nauseates adults; and to all ages it brings a similar sense of dislocation, as if the world has gone topsy-turvy, and your own sense perceptions are going haywire.

This is the power of rotation. It disorients us.

To a mathematician, this is ironic. Rotation does not scramble reality; rather, as few other actions can, it *preserves* reality. Rotation is an *isometry*, meaning that it is (along with reflection and translation) one of the few ways of moving a shape without changing its properties.

Rotation is paradoxical in this way: it disorients but does not disorganize. Perhaps that's why we are drawn to the beauty of the five-legged starfish, the six-pointed snowflake, and the seven-petaled flower. For such figures, rotation is a *symmetry*, an action that leaves the structure unchanged. Objects with rotational symmetry almost seem to spin while standing still—and to stand still while spinning.

In the pages ahead, we hope you find yourself simultaneously oriented and disoriented. These rotations will defy your rectilinear expectations—but perhaps satisfy some other expectation, the same curious expectation that draws us back to spinning carnival rides in spite of our stomachs' urgent counsel.

wobble

begin with a small black circle;
rotate and enlarge to create the next circle

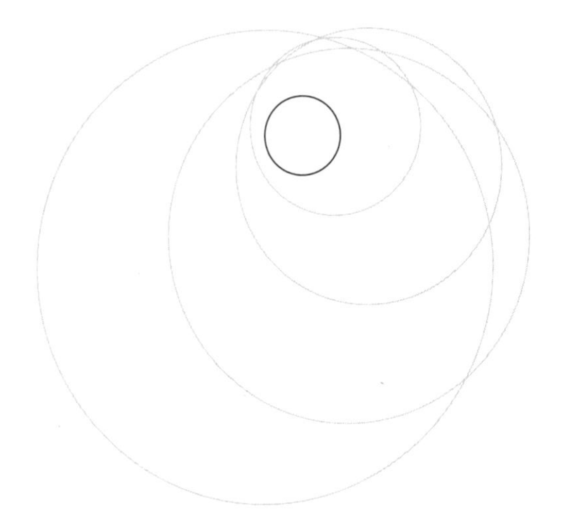

pirouette

begin with a small black square;
rotate about a circle with a radius of 0.3 inches;
origin shifted to the center of the page

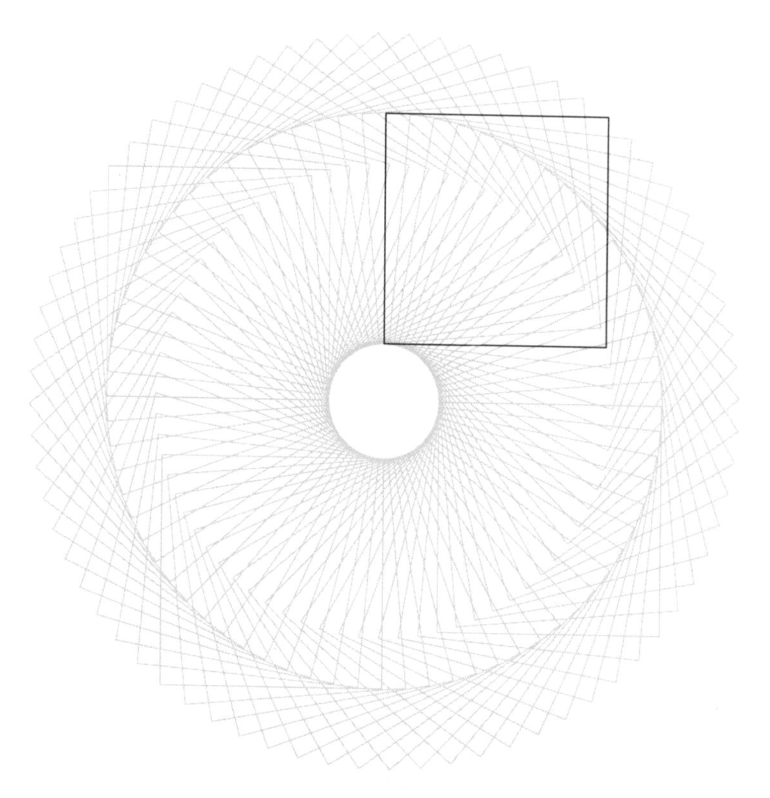

disk

begin with a small square; rotate about a circle with a radius of 1 inch; origin shifted to the center of the page

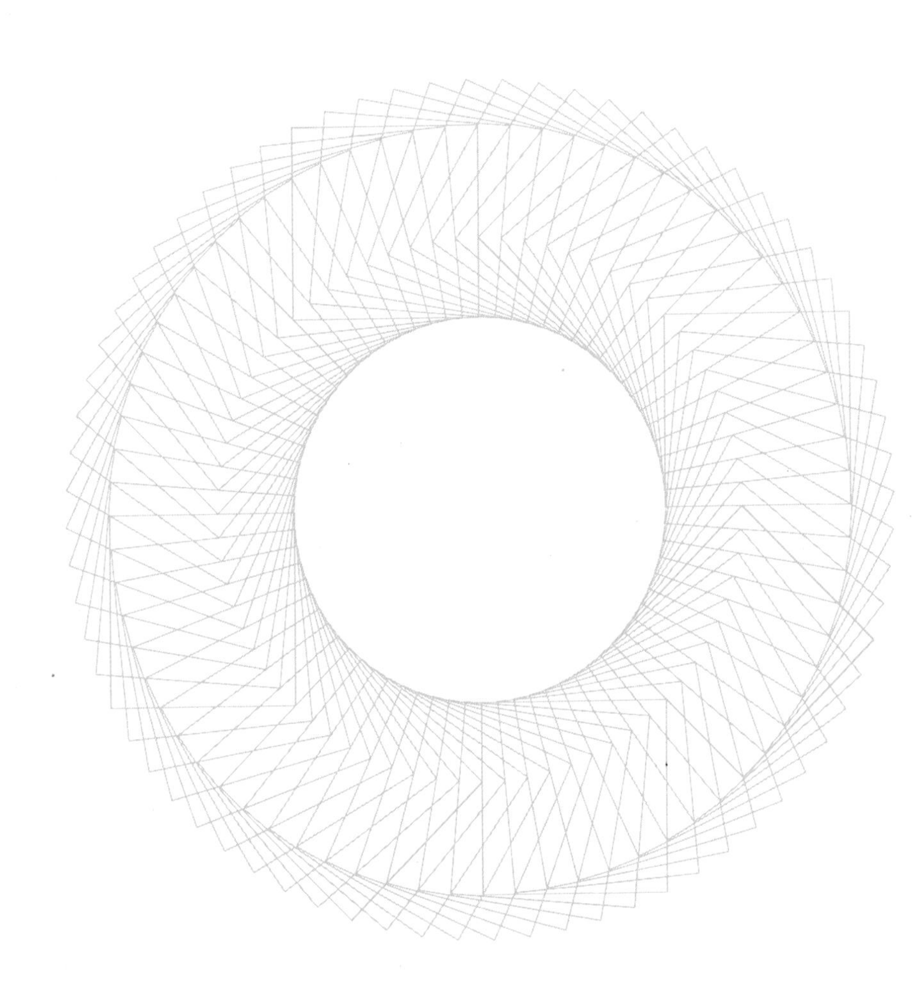

spiral staircase

begin with a small square; rotate about the center of the page with each successive square increasing in size

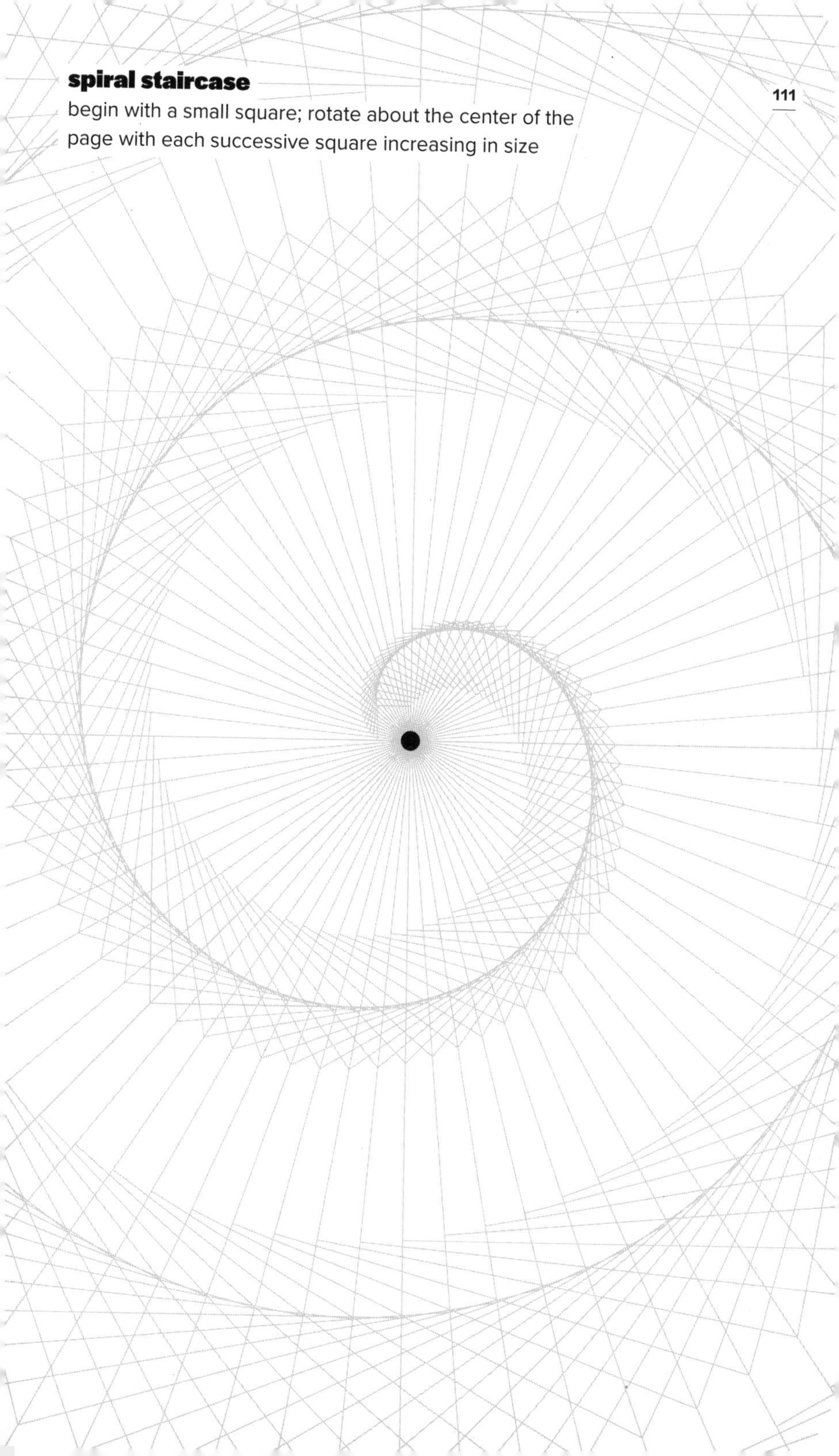

tweaked ear

rotate upper corner 90°; translate into place;
draw a white right triangle in the upper-left corner

Market Street

$x = 1$ for $y > 7.25$; $y = -(7.25/4)(x - 5)$ for $y \leq 7.25$;
rotate the page to the left of the black line 61.11°

detour

$y = c$ with a rotation of 26.5° in the middle and shifted to the left on the bottom

the four corners

rotate the center of the page 30°, 120°, 210°, and 300°

southwest

rotate about the center
of the page 135°

radiation

rotate the middle of the left edge $(30k)^\circ$ for $k = 0, 1, 2, 3, 4, 5$

half spun

$y = c$ for $-2.75 \leq x \leq 0$ and for $y > 6$;
$x = -c$ for $0 \leq y \leq 6$ and for $x < -2.75$

sideways
flip and rotate −90°
119

The artist M. C. Escher drew inspiration from the ideas of mathematics. “We do not know space,” Escher once said. “We do not see it, we do not hear it, we do not feel it. We are standing in the middle of it, we ourselves are part of it, but we know nothing about it.” Here, the rotated reptiles, in their endless repetition, offer a striking case of disorientation without disorganization.

CHAPTER: 8

Scaling the rhyme of large and small

Perhaps it goes without saying, but in physical reality, size matters.

It's not just that a large pizza feeds more people than a small one. It's that a large pizza and a small pizza exhibit wholly different properties, from cooking time to crust-to-cheese ratio. At more extreme sizes, the difference is even starker: at the vast scale of galaxies, gravity dominates, while at the tiny scale of molecules, gravity all but vanishes.

But this, we suspect, does not go without saying: in mathematical imagination, size doesn't matter.

One of math's joys is that we can scale endlessly up and down. The span from 0.3 to 0.4 has all the same properties as the span from 0.33 to 0.34, or from 0.333 to 0.334, or from 0.33333333 to 0.33333334. In mathematics, the orders of magnitude all rhyme. Things large and small possess the same fundamental structure.

Is a notebook page a physical object or a work of mathematical imagination? Both, of course. Thus this chapter plays on both conceptions of scale. We're reminded of a passage from Neil Gaiman's *American Gods*, in which a character gazes up at the night sky. "Shadow could not decide whether he was looking at a moon the size of a dollar, a foot above his head . . . [or] a moon the size of the Pacific Ocean, many thousands of miles away."

In physical reality, the two possibilities stand opposed. But in mathematics, they coincide. "Perhaps," writes Gaiman, "it was all a matter of perspective."

close-up

scale the page 2.5 times larger

scale model

scale a standard page ½ times smaller and center

two by two

scale a standard page ½ times smaller and place in a 2 × 2 grid

compressed

shrink the vertical scale by 1/40th the original scale

vanishing

starting with a standard page, shrink the current page by 80%, position in the middle of the page; repeat

stacked

starting with a standard page, shrink the current page by 70% and rotate 180°; position at top left; repeat

mosaic, 15 columns

mosaic with 15 columns of circles

mosaic, 20 columns

mosaic with 20 columns of circles

mosaic, 33 columns

mosaic with 33 columns of circles

mosaic, 42 columns

mosaic with 42 columns of circles

$y = c$; for each line use line thicknesses of 0.5, 1, 3, and 5 times the thickness of a standard line, all with opacity/transparency of 50%

Scaling is about more than just size. It can also operate as a kind of metaphor for any gradual process. One example is transparency. Some objects are opaque, obscuring whatever is behind them; other objects are transparent, which is to say, invisible; and between these extremes are the various degrees of opacity and transparency. To interpolate between the extremes is a sort of nonliteral scaling.

CHAPTER: 9

Polar the world of the compass rose

"Civilized man," wrote mathematics popularizer Martin Gardner, "is surrounded on all sides, indoors and out, by a subtle, seldom-noticed conflict between two ancient ways of shaping things: the orthogonal and the round."

Thus far we have taken the orthogonal perspective. That's because—even while creating round figures such as parabolas, circles, and waves—we have used rectangular coordinates. This system locates every point on a rectilinear grid of verticals and horizontals.

But there is another way: polar coordinates. First, imagine a compass rose somewhere on the page. Eastward (that is, rightward) shall be called zero. Moving counterclockwise, northward is 90°, westward is 180°, southward is 270°, and as the angle keeps increasing, the cycle repeats, so that 360° again points eastward.

Now, to describe a point, we give two numbers: first, its angle (θ) on the compass rose, and second, its distance (often written r) from the center of the compass rose. For example, if the compass rose is in the middle of a recto page, then the top center is at an angle of 90° and a distance of 4.25. Meanwhile, the center of the spine is at an angle of 180° and a distance of 2.75.

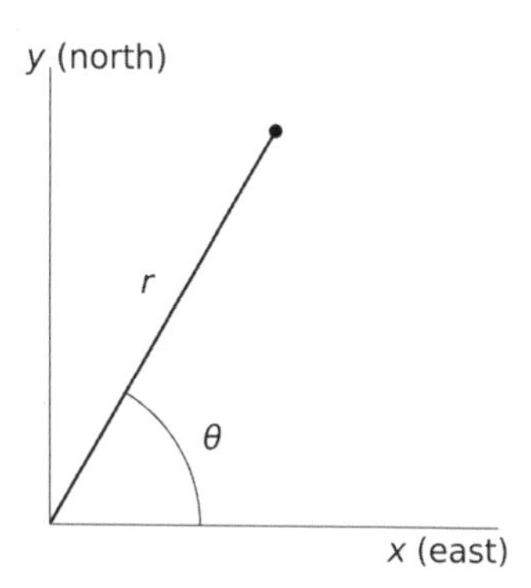

This new language is perfect for spirals, loops, and other rounded figures that previously eluded our grasp. More than a new collection of curves, this chapter marks a new conception of space itself.

sunshine

$\theta = c$ for $\pi \leq \theta \leq 3\pi/2$; origin shifted to (0, 7.25)

clover $r = \cos(4\theta) + \sin^2(4\theta) + c$ for $0 \leq \theta \leq 2\pi$; origin shifted to the center of the page

unspooled

$y = c$; vertical spiral is a lituus spiral with polar coordinates $r = \sqrt{\frac{1}{\frac{\pi}{2} - \theta}}$ for $-25\pi \leq \theta \leq 2$, origin shifted to the center of the page

galaxy

$r = \sqrt{\theta}$ and $r = -\sqrt{\theta}$ for $0 \leq \theta \leq 12\pi$; origin shifted to (2.75, 3.4)

torpedo

$r = c \cos(\theta) \cos(2\theta)$ for $0 \leq \theta \leq \pi$, rotated and shifted on the page

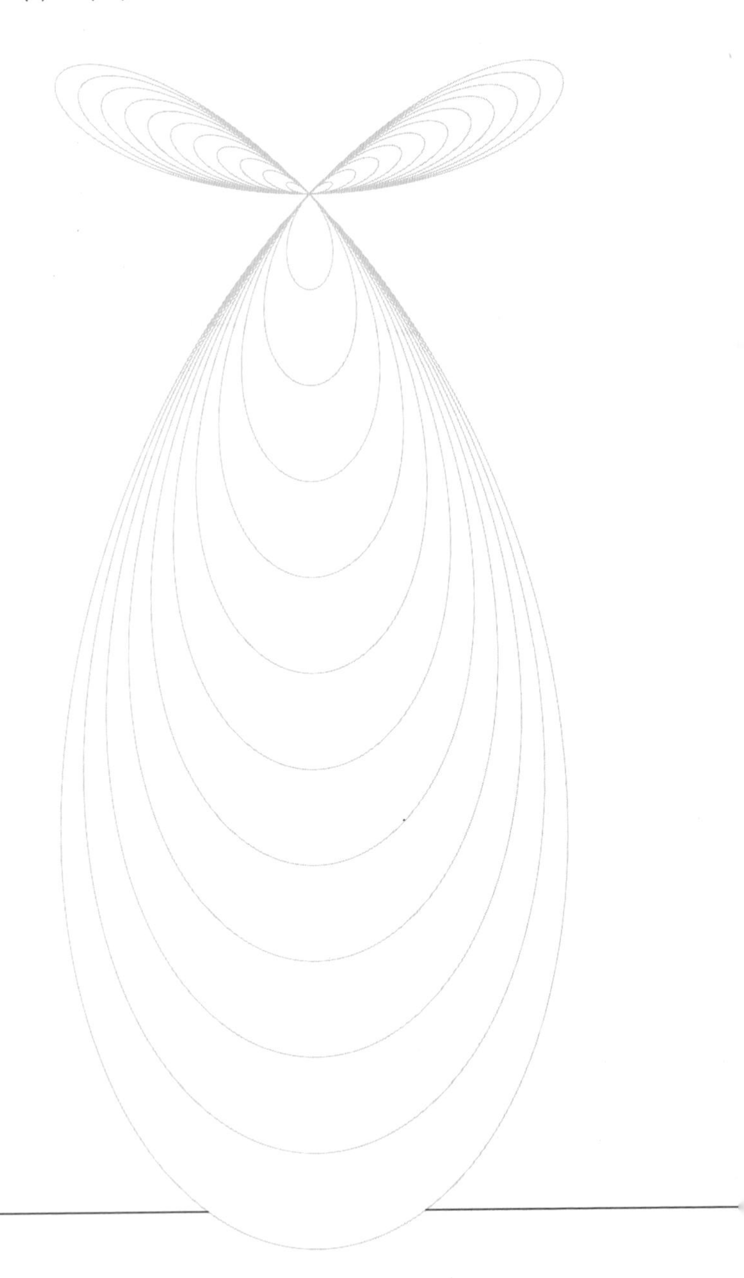

hidden spiral

$y = c$; overwrite with thick white spiral $r = \theta$ for $0 \le \theta \le 24\pi$, origin shifted to the center of the page

daisy

$r = 4.25 \cos(12\theta)$ for $0 \leq \theta \leq 2\pi$;
origin shifted to the center of the spine

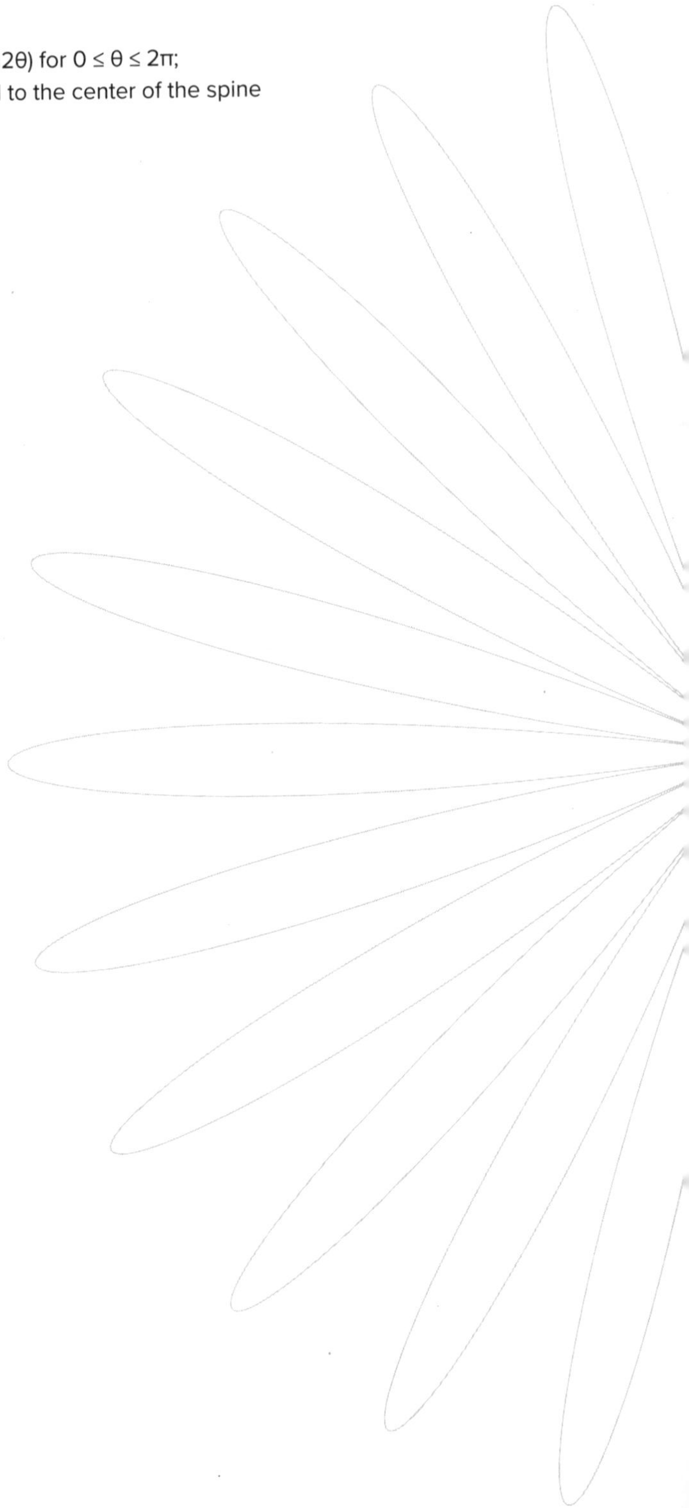

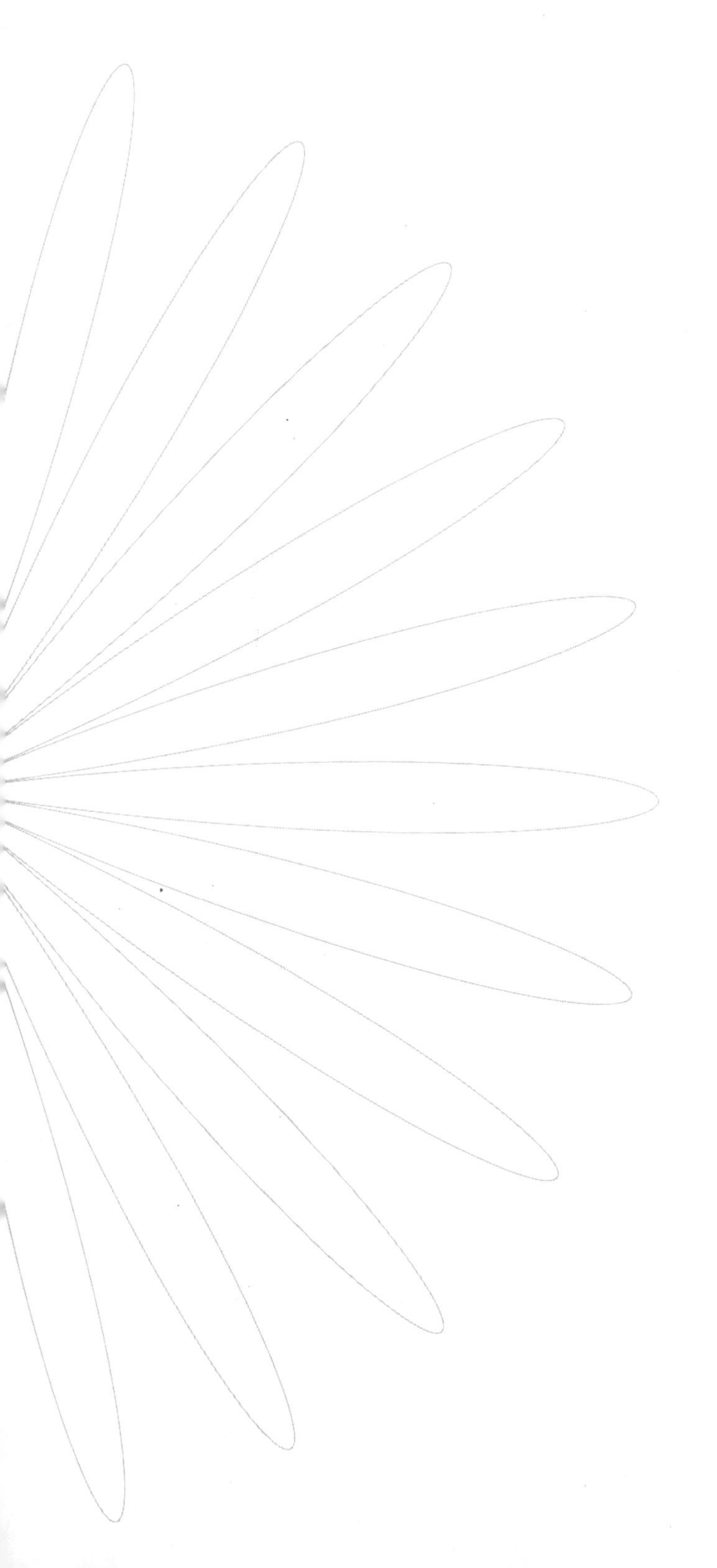

12 stitches

mark every 30 degrees around the circle

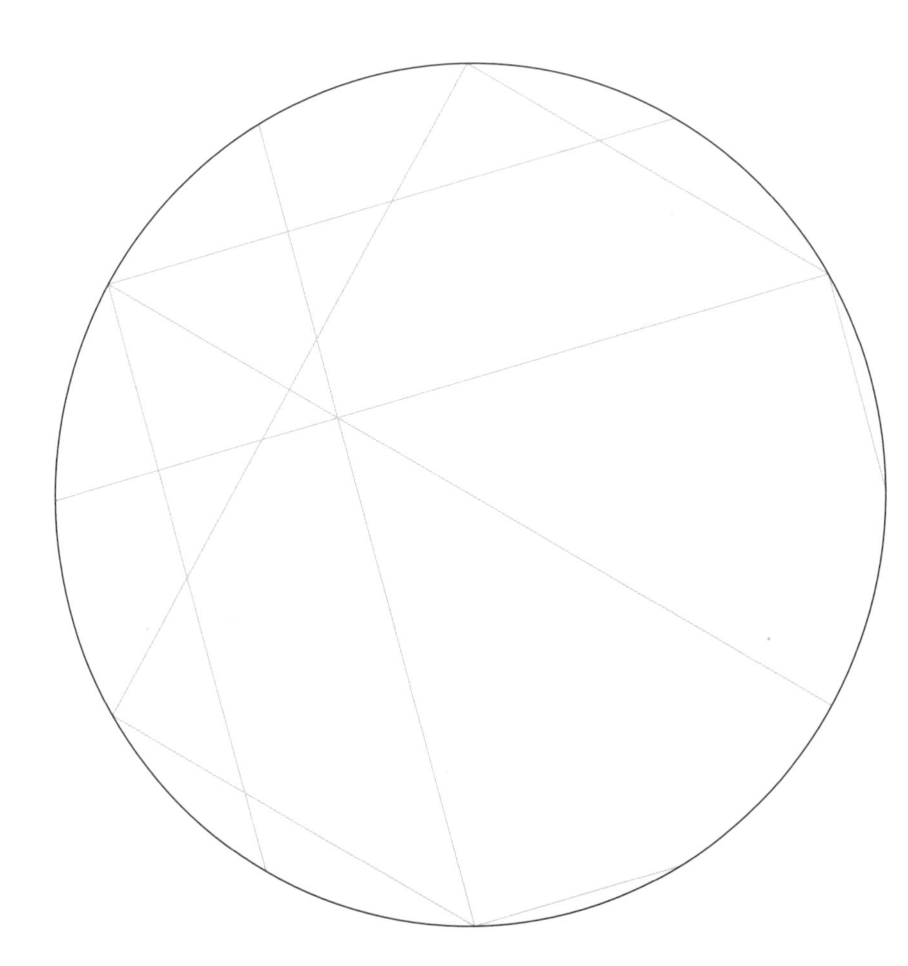

To create this page, draw a circle. Then make a mark every 30 degrees around the circle. Number those marks from 1 to 12. Draw a line from mark 1 to mark 2. Draw a line from mark 2 to mark 4. As a general rule, draw a line from mark k to mark $2k$. (Note, mark 10 will be connected to mark 20. So, mark 1 is also mark 13, mark 2 is also mark 14, and so forth.) On the next four pages we increase the number of marks around the circle. This process is called curve stitching.

24 stitches

mark every 15 degrees around the circle

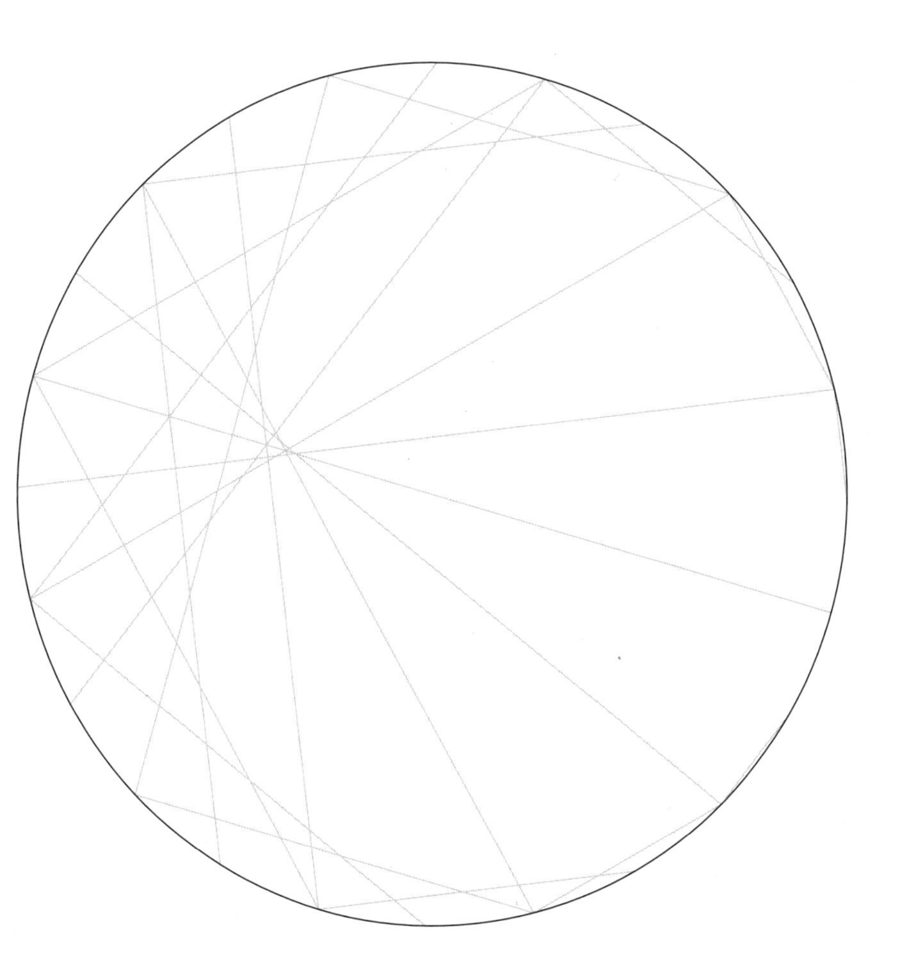

72 stitches

mark every 5 degrees around the circle

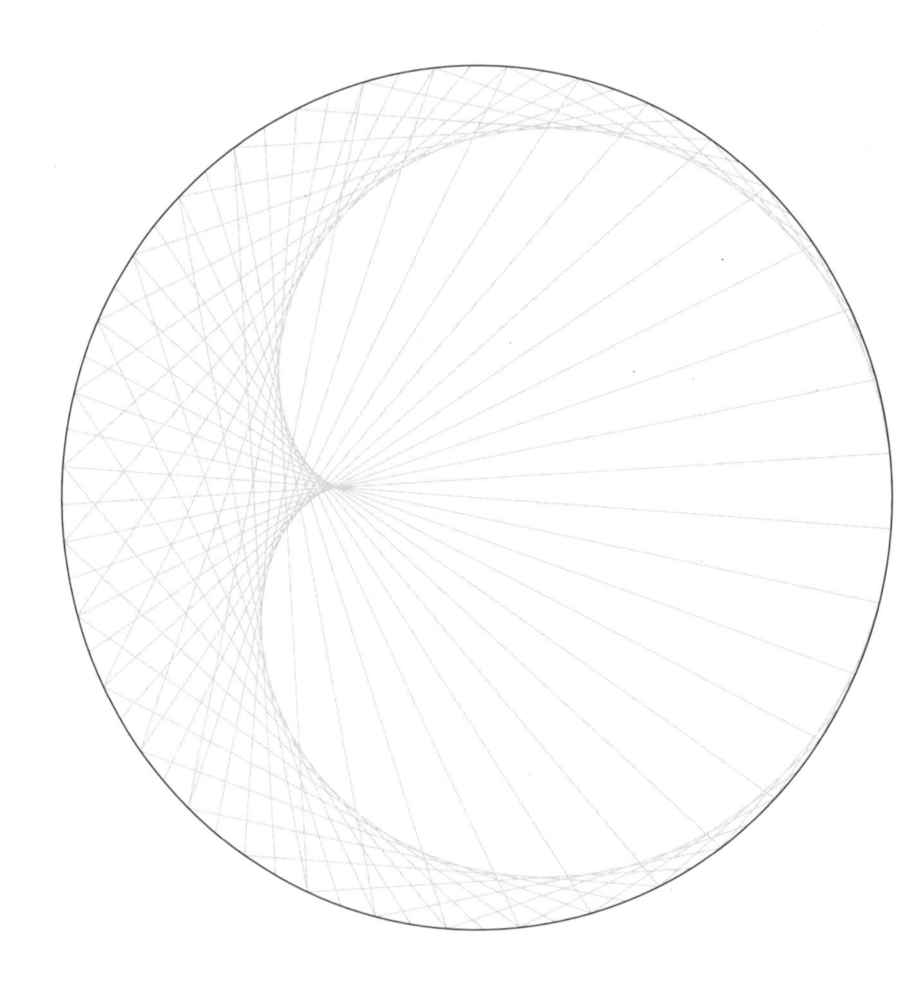

Do you see the shape emerging? Look at the next page.

cardioid

$r = c[1 + \cos(\theta)]$; origin shifted up and slightly to the right

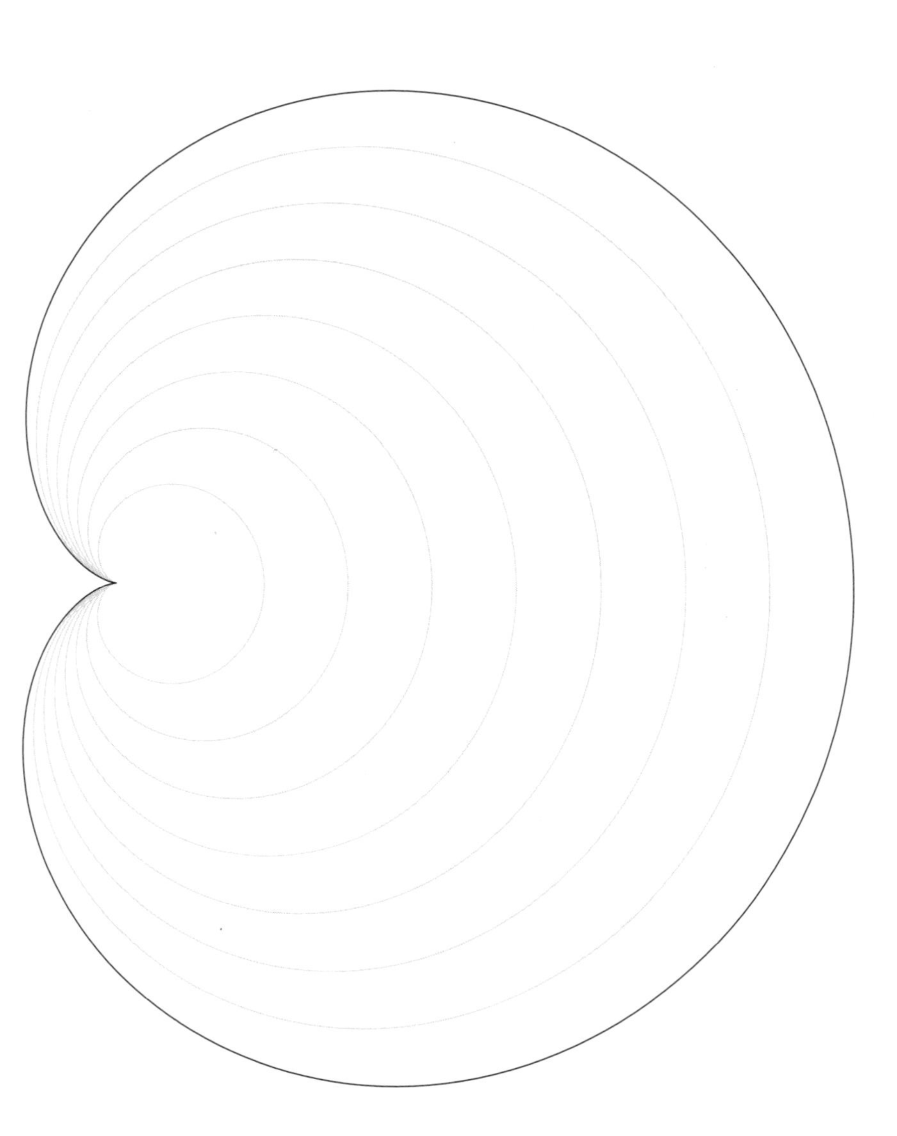

The shape that has emerged is called a cardioid. Several cardioids of various sizes are drawn on this page. The shape can be created with polar coordinates using $r = 1 + \cos(\theta)$ for $0 \leq \theta \leq 2\pi$.

propeller

$r = -\theta \cos(3\theta)$ for $-3\pi - \pi/2 \leq \theta \leq 14\pi - \pi/2$, scaled to fit the page and shifted over and up

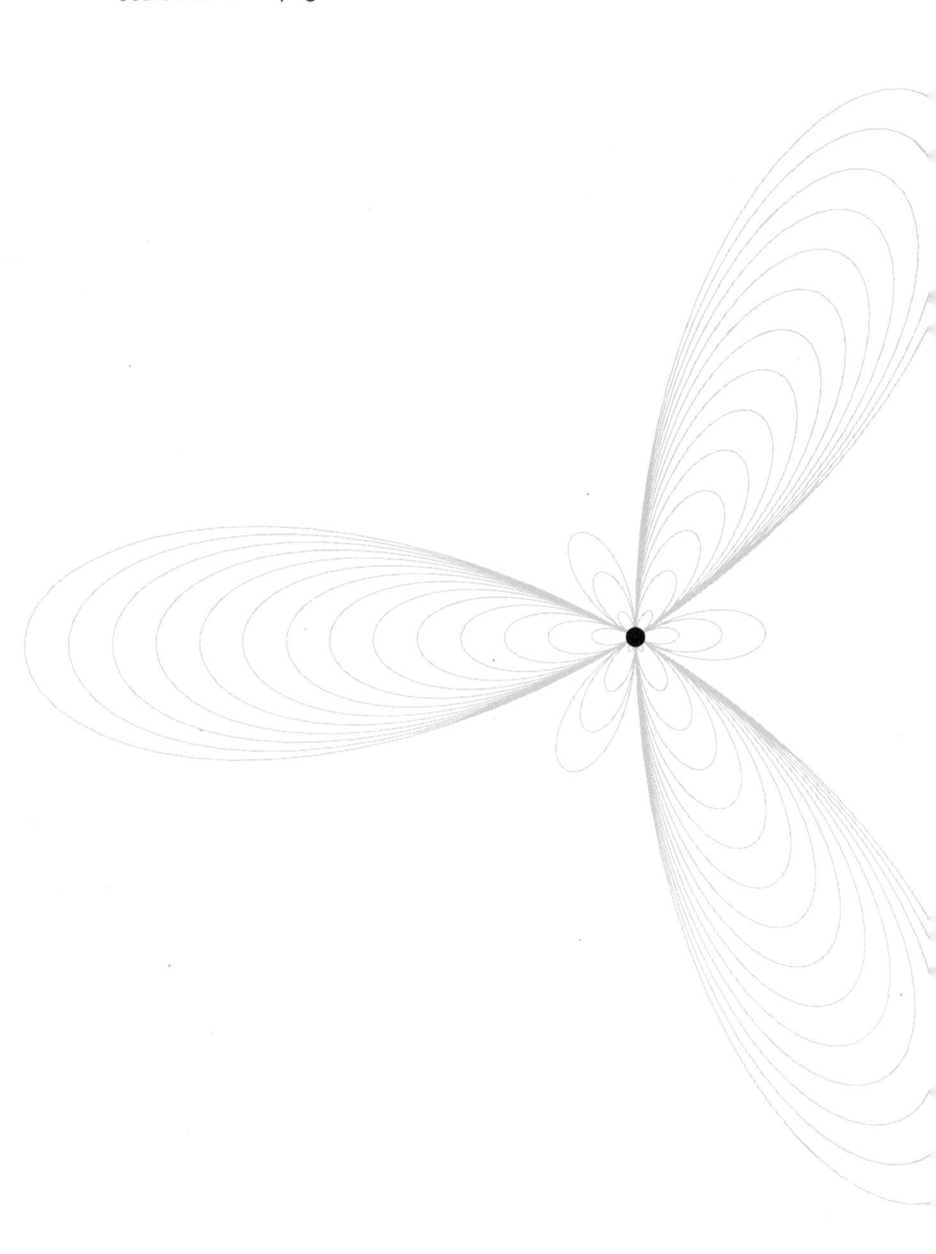

corsage

$r = \theta + 2\sin(2\pi\theta)$ for $0 \le \theta \le 14\pi - \pi/2$;
origin shifted to the center of the page

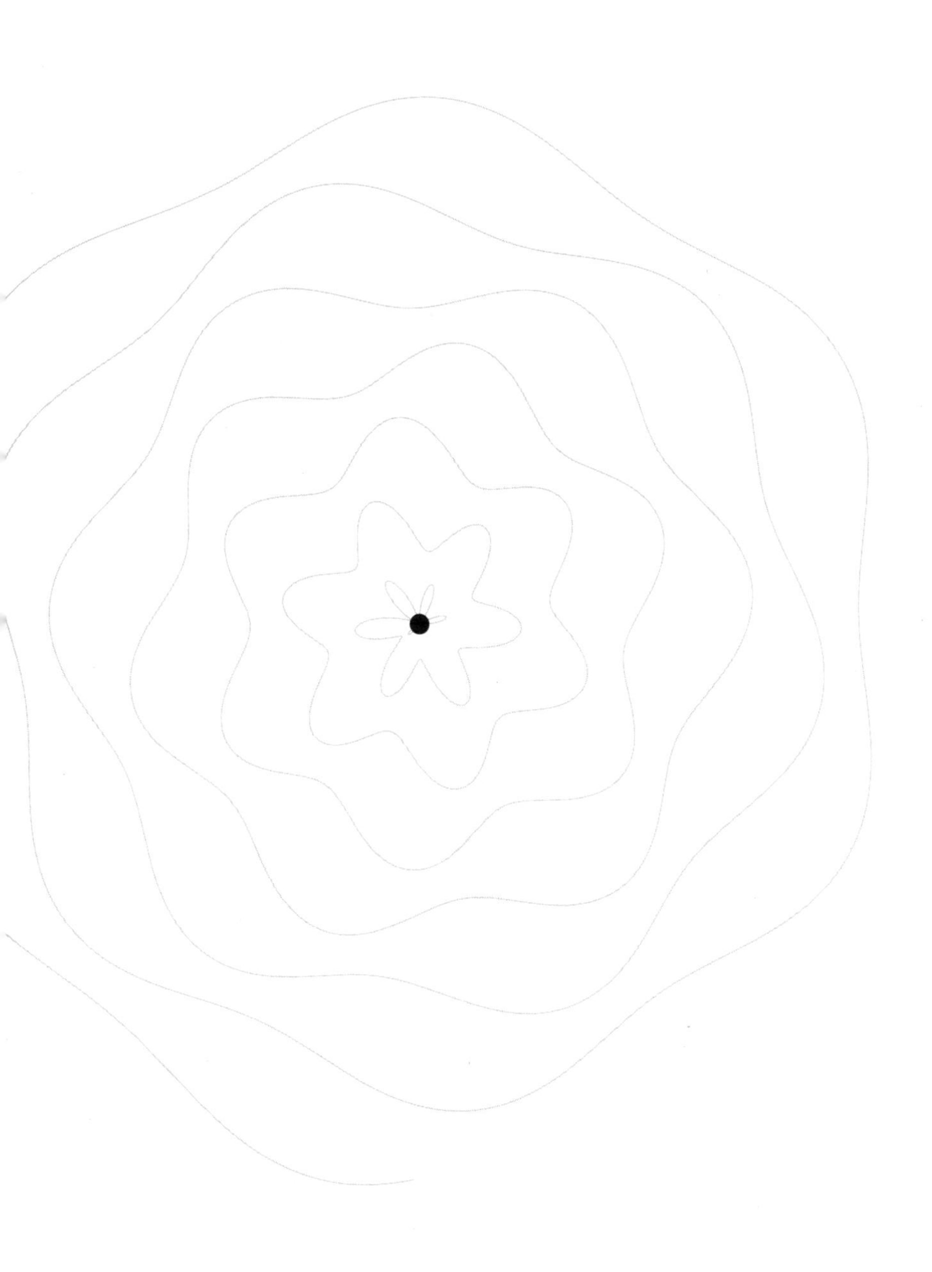

autumn

$y = c$; fill the curve
$r = 5.2 \cos[\sin(4\theta)]$
$+ 0.15 \sin(80\theta)$ with white;
scale, rotate, and translate
four copies

mandala

variations on the polar curves $r = 2 + \frac{0.5\cos(12\theta)}{1.5+|\sin(12\theta)|}$, $r = 1.5 + 0.25\cos(6\theta)$, and $r = 1.25$ for $0 \le \theta \le 2\pi$; origin shifted to the center of the page

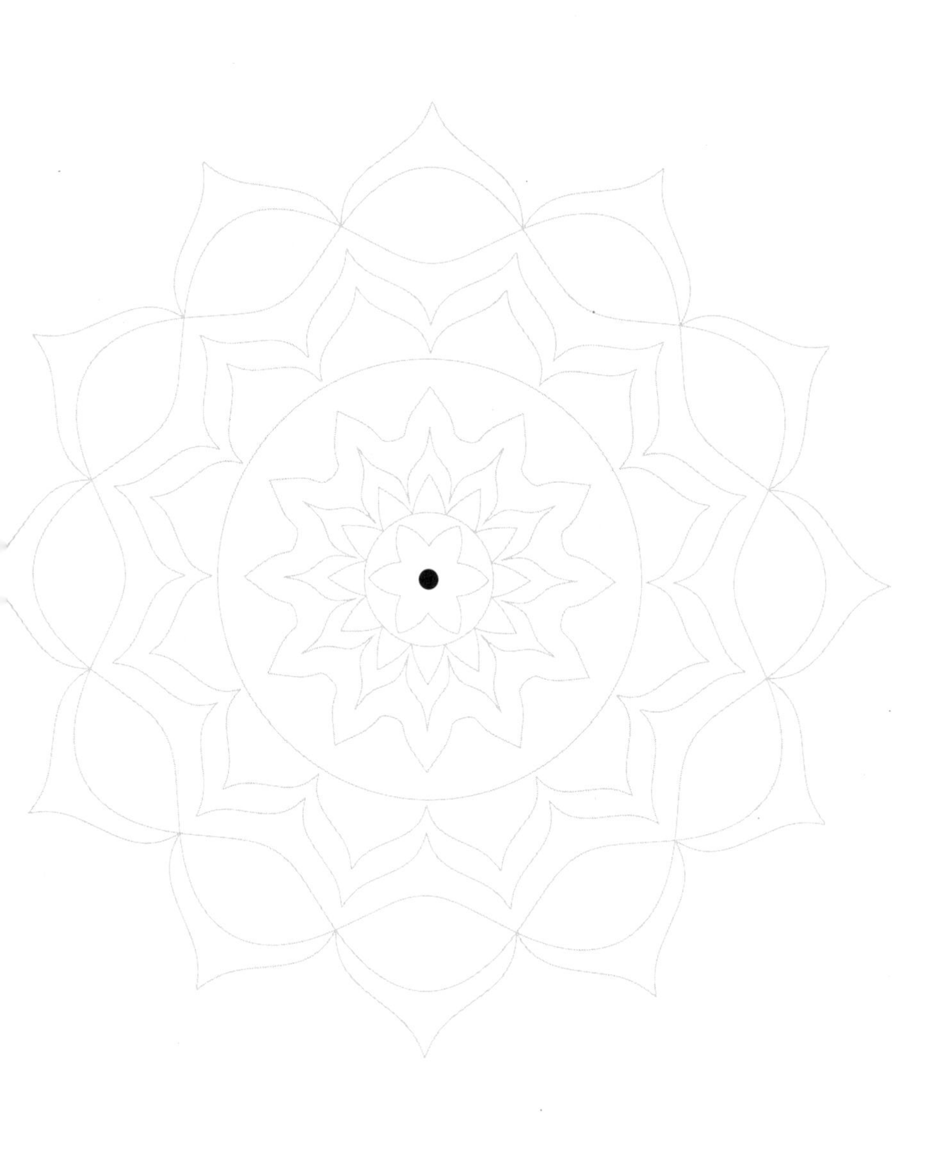

Folium de Dürer

$r = c \sin(\theta/2)$ for $0 \leq \theta \leq 4\pi$; origin shifted to the center of the page

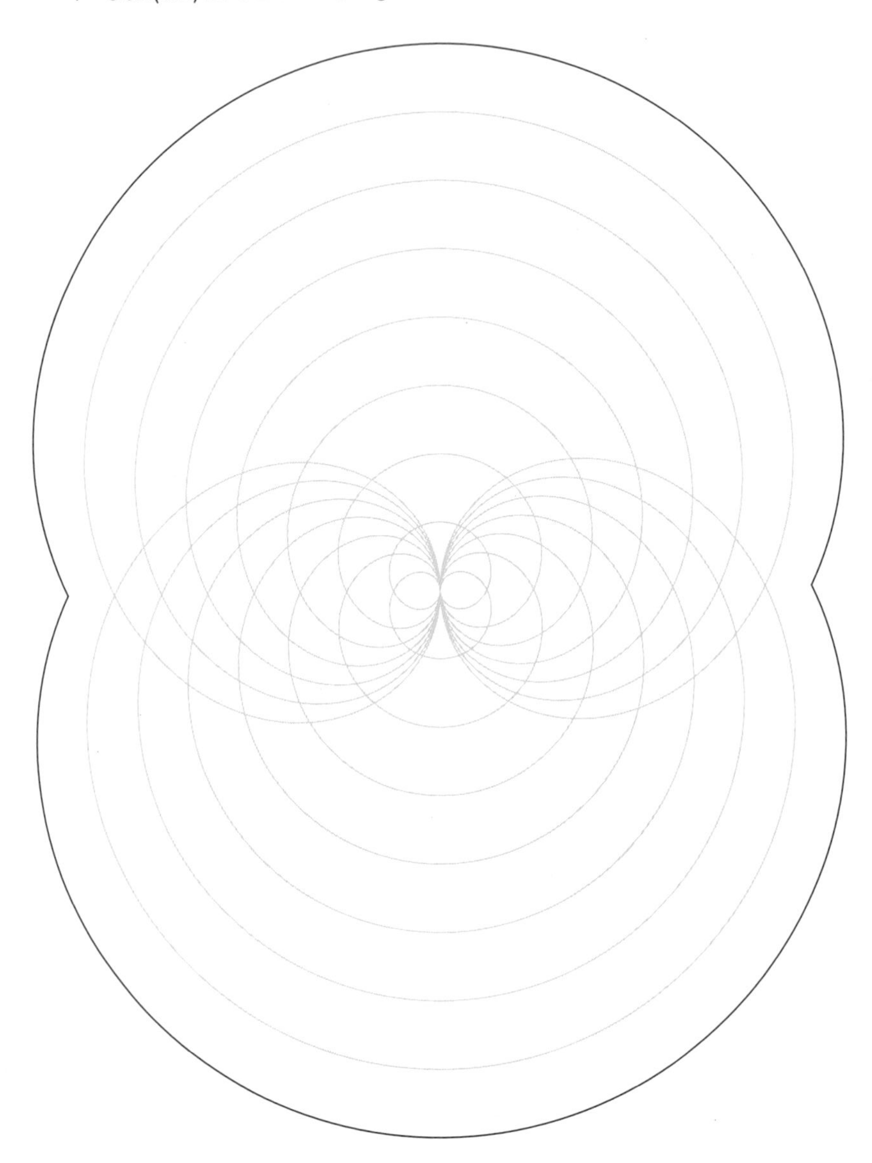

Named for the 15th-century German artist Albrecht Dürer, this lovely, looping curve is a particular specimen of the species known as the rose. But Dürer's folium is no ordinary rose: it can be employed to help split an angle into three equal parts, an act known as trisecting the angle. Such trisection is impossible with the traditional tools of geometry—which goes to show that polar coordinates are no traditional tool. The polar perspective opens up possibilities of which rectangular coordinates can only dream.

CHAPTER: 10

Pathways parametric footprints

All the curves in this book (aside from those in the previous chapter) were born from x-y equations. Each curve embodies a relationship between x and y, the two variables inseparably paired.

But what if x and y could be teased apart?

Picture a fly crawling around on a wall. Instead of writing a single equation to describe its path, we can write two parametric equations. At each moment in time, the first equation tells us the fly's x coordinate, and the second tells us its y coordinate. To discover the fly's location at a particular moment (say, noon), simply use the equations in tandem.

Parametric equations are a wildly flexible language. Divorcing x from y gives them both unprecedented freedom and allows them to collaborate on shapes of exquisite complexity. The orthogonal figures of rectangular coordinates, the rounded curves of polar coordinates, scribbles ranging from zigzags to lazy loops—all these pathways (and more) are encompassed by the parametric approach.

Why "parametric"? Because the key is the *parameter*. Throughout this book, we have used the parameter c to describe many curves at once. Now, we use the parameter t to describe many moments—that is, many points—at once. A parameter is a sort of mathematical amphibian, halfway between a constant and a variable: a constant that varies, or a variable that stays constant.

In any case, as you will see on the pages to come, the parameter unlocks extraordinary diversity. Parametric equations trace out all the pathways of our imagination—and even some pathways beyond imagining.

bicycle race

$x = t - \sin(t); y = 1 - \cos(t) + c$

butterfly

$x = \sin(t)\left[e^{\cos(t)} - 2\cos(4t) - \sin^5(t/12)\right]$ and
$y = \cos(t)\left[e^{\cos(t)} - 2\cos(4t) - \sin^5(t/12)\right]$
for $0 \le t \le 12\pi$, shifted up and to the right

passageway

$x = \cos(10t)\sin(10t)$ and $y = \sin(20t)\cos(7t)\sin(20t)$
for $0 \leq t \leq 2\pi$, shifted up and to the left

slinky

$x = 4\pi t + 4\pi[\cos(80\pi t) - \cos(2\pi t)]$ and $y = \cos(4\pi t) + 4\pi[\sin(80\pi t) - \sin(2\pi t)]$ for $0 \leq t \leq 1$, shrunk and shifted up and over

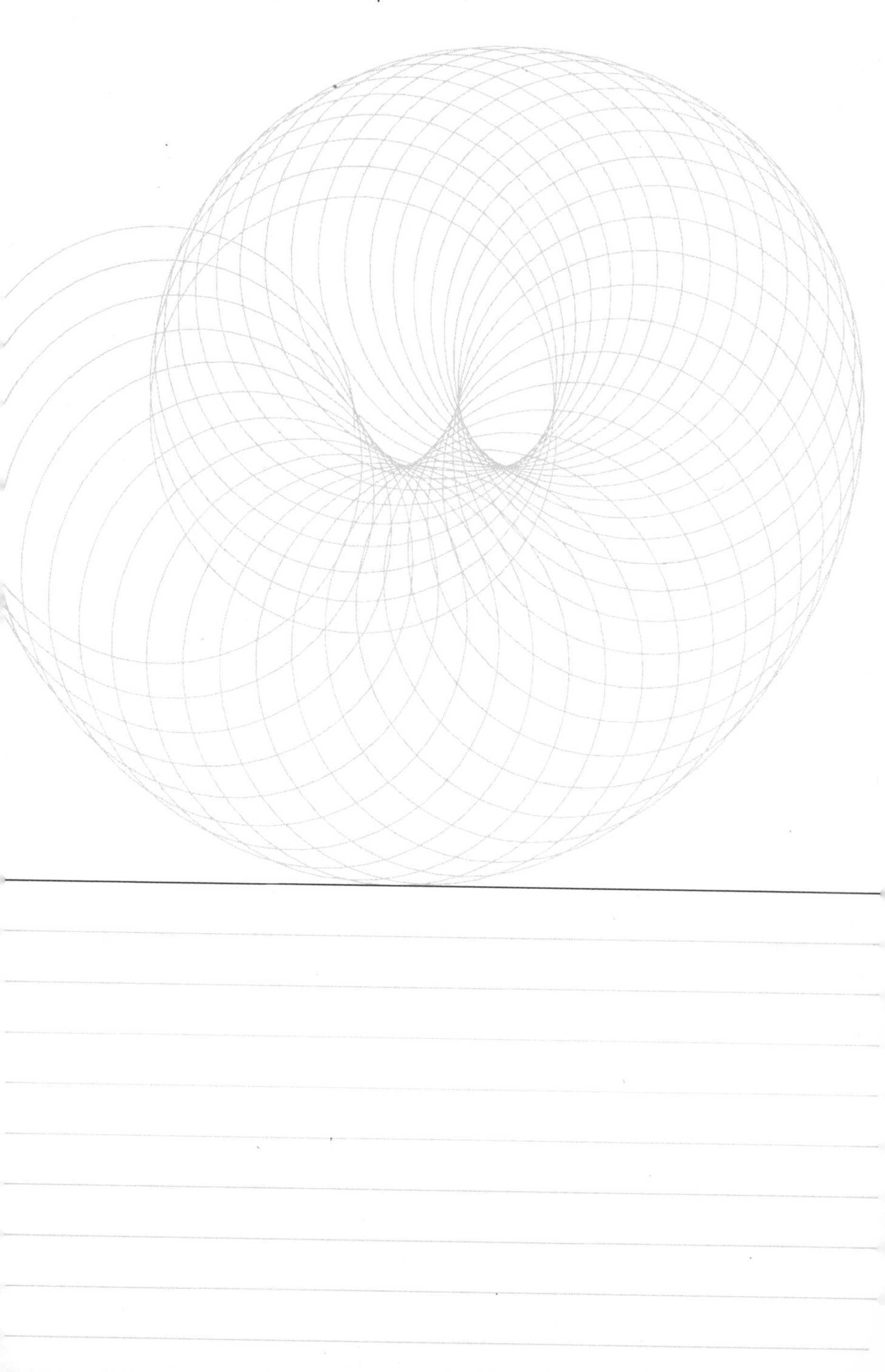

valentine

$x = 16 \sin^3(t)$ and $y = 13 \cos(t) - 5 \cos(2t) - 2 \cos(3t) - \cos(4t)$ for $0 \le t \le 2\pi$, shifted up and to the left and scaled to various sizes

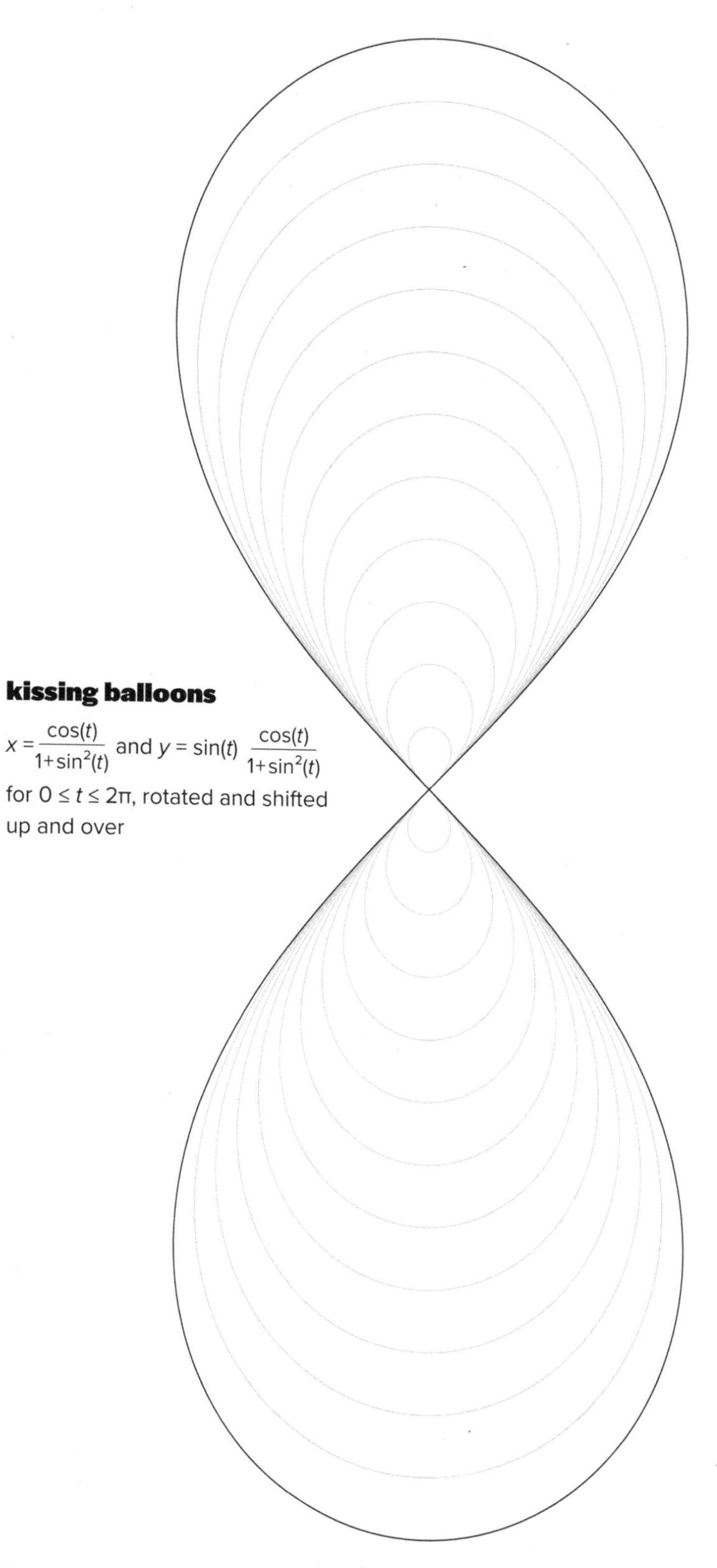

kissing balloons

$x = \dfrac{\cos(t)}{1+\sin^2(t)}$ and $y = \sin(t)\ \dfrac{\cos(t)}{1+\sin^2(t)}$

for $0 \le t \le 2\pi$, rotated and shifted up and over

wormhole

$x = 2[\cos^2(6.1t)]\,\sin[\sin(7.6t)]$ and $y = 2[\sin(6.1t)]\,\cos[\cos(7.6t)]$ for $-8 \leq t \leq 8$; origin shifted to the center of the page

Parametric equations can create very curvy curves because, unlike traditional functions $y = f(x)$, they are not restricted by the vertical line test, which requires that any vertical line drawn through the function touch it only once. Some tests are worth failing. The parametric curvy curve on this page dramatically fails the vertical line test and is beautifully not functional.

lovelorn

$y = c$; overwrite with a white heart formed by $x = \frac{16 \sin^3(t)}{6.5} + \frac{5.5}{2}$ and $y = \frac{13 \cos(t) - 5 \cos(2t) - 2 \cos(3t) - \cos(4t)}{6.5} + 4$ for $0 \leq t \leq 2\pi$

epicycloid

$a = \frac{1}{4.5+c}$, $b = \frac{10}{4.5+c}$, $x = (a+b)\cos(t) - a\cos\left(\frac{t(a+b)}{a}\right) - \frac{5.5}{2}$,
and $y = (a+b)\sin(t) - a\sin\left(\frac{t(a+b)}{a}\right) + \frac{8.5}{2}$ for $0 \leq t \leq 2\pi$

looping

$x = t + \cos(14t)/t$ and $y = t + \sin(14t)/t$ for $0 \le t \le 20$, shifted up and to the right; $y = x$

lace 1

$x = 2\sin(13\pi t)$ and $y = 3.5\cos(\pi t)$ for $0 \leq t \leq 2$, shifted over and up

lace 2

$x = 2\ \sin(13\pi t)$ and $y = 3.5\ \cos(3\pi t)$ for $0 \le t \le 2$, shifted over and up

lace 3

$x = 2\sin(13\pi t)$ and $y = 3.5\cos(5\pi t)$ for $0 \leq t \leq 2$, shifted over and up

lace 4

$x = 2\sin(13\pi t)$ and $y = 3.5\cos(7\pi t)$ for $0 \leq t \leq 2$, shifted over and up

harmonograph

$x = -\cos(2.01t - \pi/2)e^{-0.00085t} - \cos(3t - \pi/16)$ and $y = -\sin(3t)e^{-0.0065t} - \sin(2t)$ for $0 \leq t \leq 250$; origin shifted to the center of the page

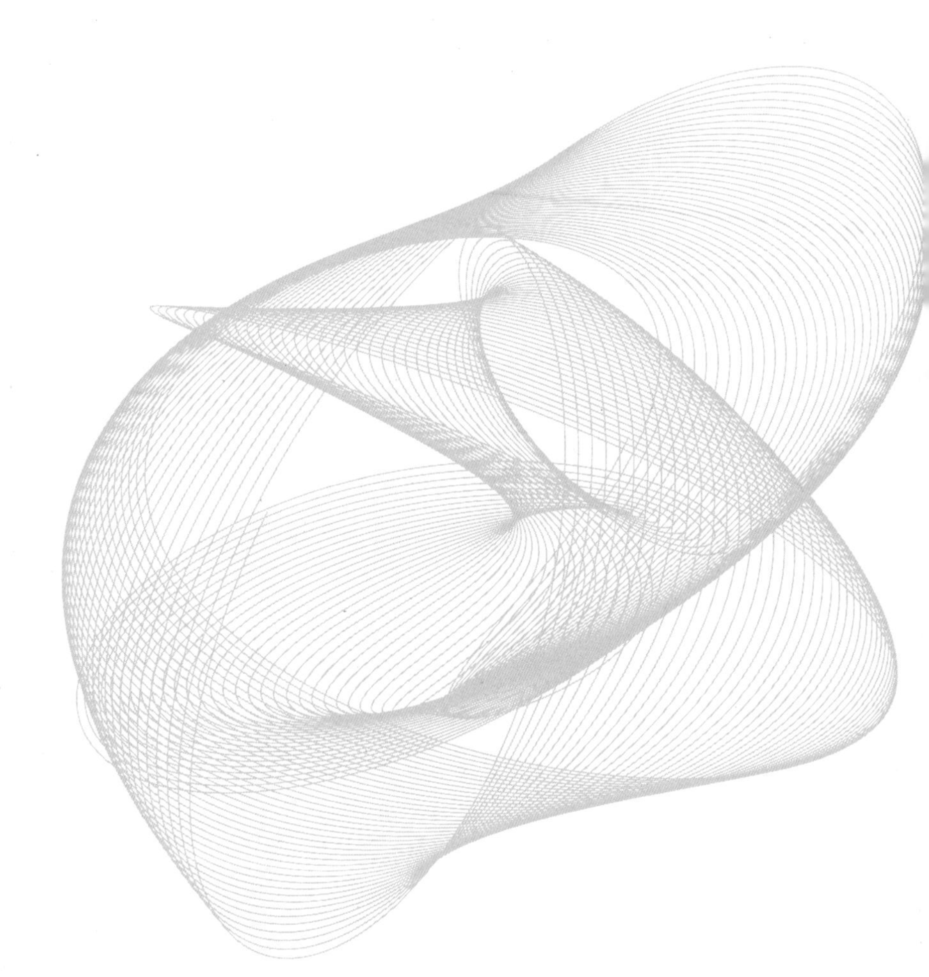

In the parametric approach, we write separate equations for x and y. Put another way, we divorce horizontal motion from vertical motion. That is precisely the logic of the *harmonograph*, a drawing device popular in the late 1800s. Two pendulums control the movement of a pen, one its horizontal motion and one its vertical motion. The result is a variety of beautiful curves: the fruits of liberating x from y and y from x.

CHAPTER: 11

Randomness serendipity through chaos

Apophenia is the human tendency to see patterns in everything, no matter how patternless. In scattered stars, we find constellations. In a noisy din, we hear names and whispers. In disconnected events, we see vast conspiracies. True randomness, it seems, is the one thing that our pattern-seeking minds can never accept. It's our blind spot.

And, lucky for us, it is also our superpower.

The human brain is a kind of engine that transforms randomness into meaning. Pour a little chaos into the gas tank, then watch as the mind zooms forward, propelled by this unlikely fuel. Many musicians and artists exploit this phenomenon by drawing creative inspiration from prompts on randomly shuffled cards. The process echoes ancient rituals in which the cracks in a burning bone or the path of a wild bird would help hunters determine where to seek their quarry.

Though true randomness is hard to achieve, mathematicians are good at generating pseudorandom numbers. They emerge from fixed equations yet are effectively unpredictable. See if these flights of mathematical randomness can prompt some apophenia of your own.

random lines

connect $(0, c)$ and a random point on the left-hand side of the page

black holes

create 3 circles; let $y = c/2$ where lines do not overlap circles; draw vertical lines when a line intersects a circle

messy desk

rotate about a random point on the page by a random number of degrees

silenced

$y = \sin(2x) + c$ for $0 \leq x \leq 1.5 \sin(2c) + 3.5$

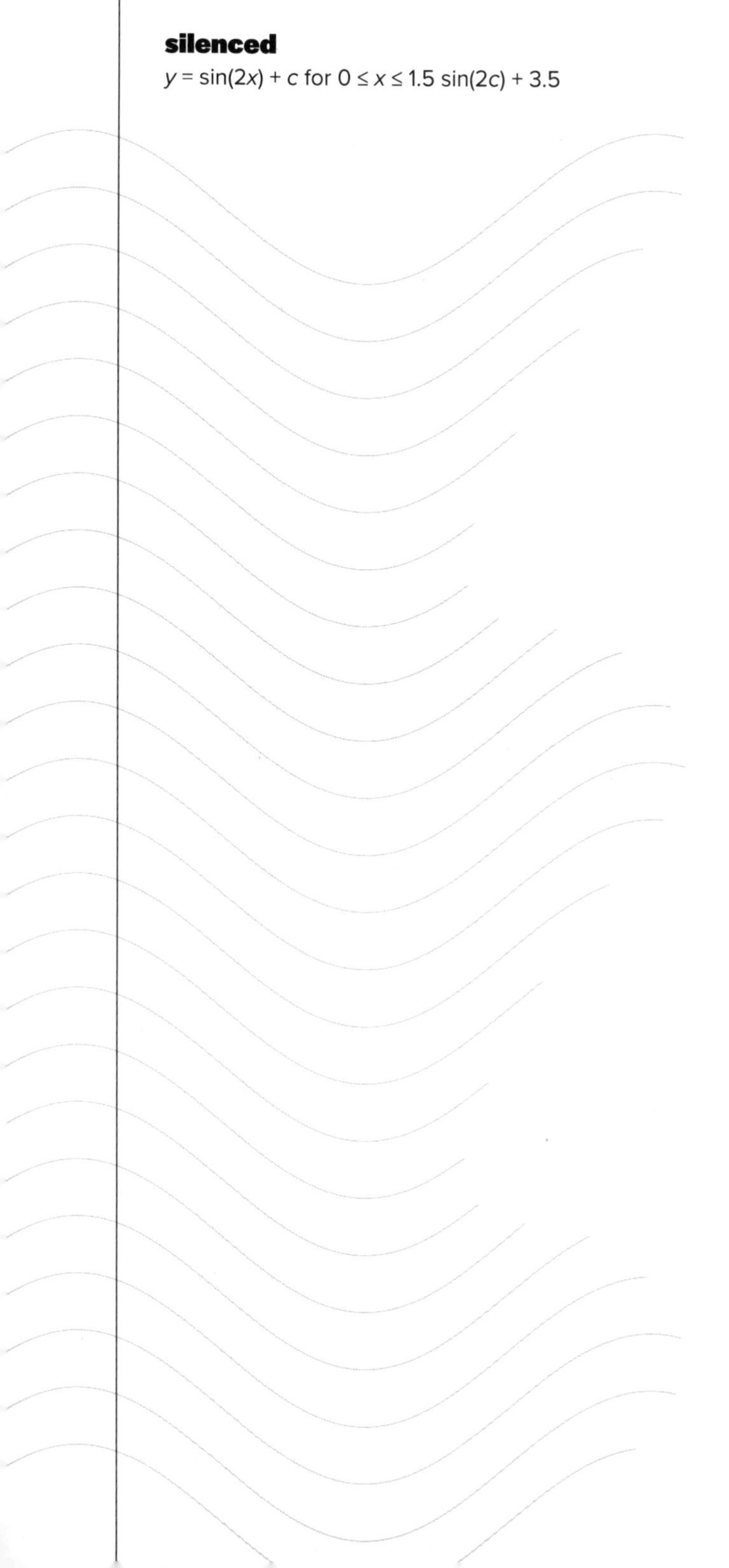

threads

$y = mx + c$, where m is a random number between −0.3 and 0.3

cubist

take a random rectangular region of the page;
rotate randomly 90°, 180°, or 270°

melting lines

$y = c$; melting lines created with curve of the form $y = e^{0.2x} \sin(-2x) + \sin(-3x)$

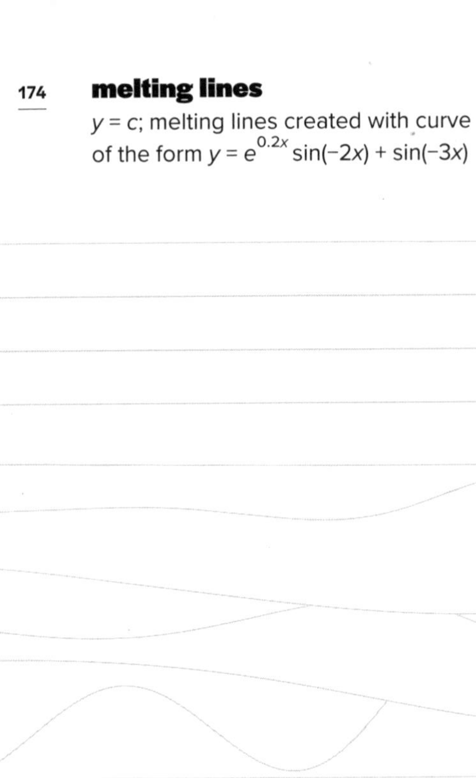

ink runs

$y = c$; pick random points along the line and fill between $y = \sin(x)$ and $y = -\sin(x)$, stretched over line segments of random length between 0.1 and 0.4

falling lines

the fallen lines connect the points (0, c) and ($-b$, 0) or the points (-5.5, c) and ($-5.5 + b$, 0) for $b = 5.5 \sin\left[\cos^{-1}\left(\frac{c}{5.5}\right)\right]$

random waves

$y = 0.1 \sin(10x) + 0.05 \sin(80kx) + c$,

where k is a random number between 0 and 1

178

random up-down

draw 12 random horizontal lines and 13 random vertical lines;
draw a black line at $y = 7.25$

CHAPTER: 12

Third Dimension the paradox of paper

Human vision is the solution to an impossible riddle.

You see, the world has three dimensions: length, width, and depth. But our eye gathers information on a two-dimensional retina: it has length and width, but no depth. Thus, from an incomplete set of data, the brain must reconstruct a three-dimensional reality.

The process is a miracle—but an inherently imperfect one. That's why we fall prey to optical illusions. Clever artists can exploit our brain's shortcuts by feeding it clues that contradict one another or that conjure false images. It's nothing to be ashamed of—you might say our vision itself is a kind of happy illusion.

In this final chapter of the book, we attempt some optical illusions of our own. We seek to create 3D experiences with 2D paper. In doing so, we follow in the footsteps of artist Jeff Nishinaka. "I feel that paper has energy and a life of its own," he once said, "and that I am only releasing or revealing what is already there."

around the corner

draw a line connecting the points (5, 8.5) (which is off the page) and (−1, c) for $-1 \le x \le 0$; draw a line connecting the points (−1, c) and (−13.5, 8.5) (also off the page) for $-5.5 \le x \le -1$

bowl

$x = (y - c)^2 - c$, shifted to the right

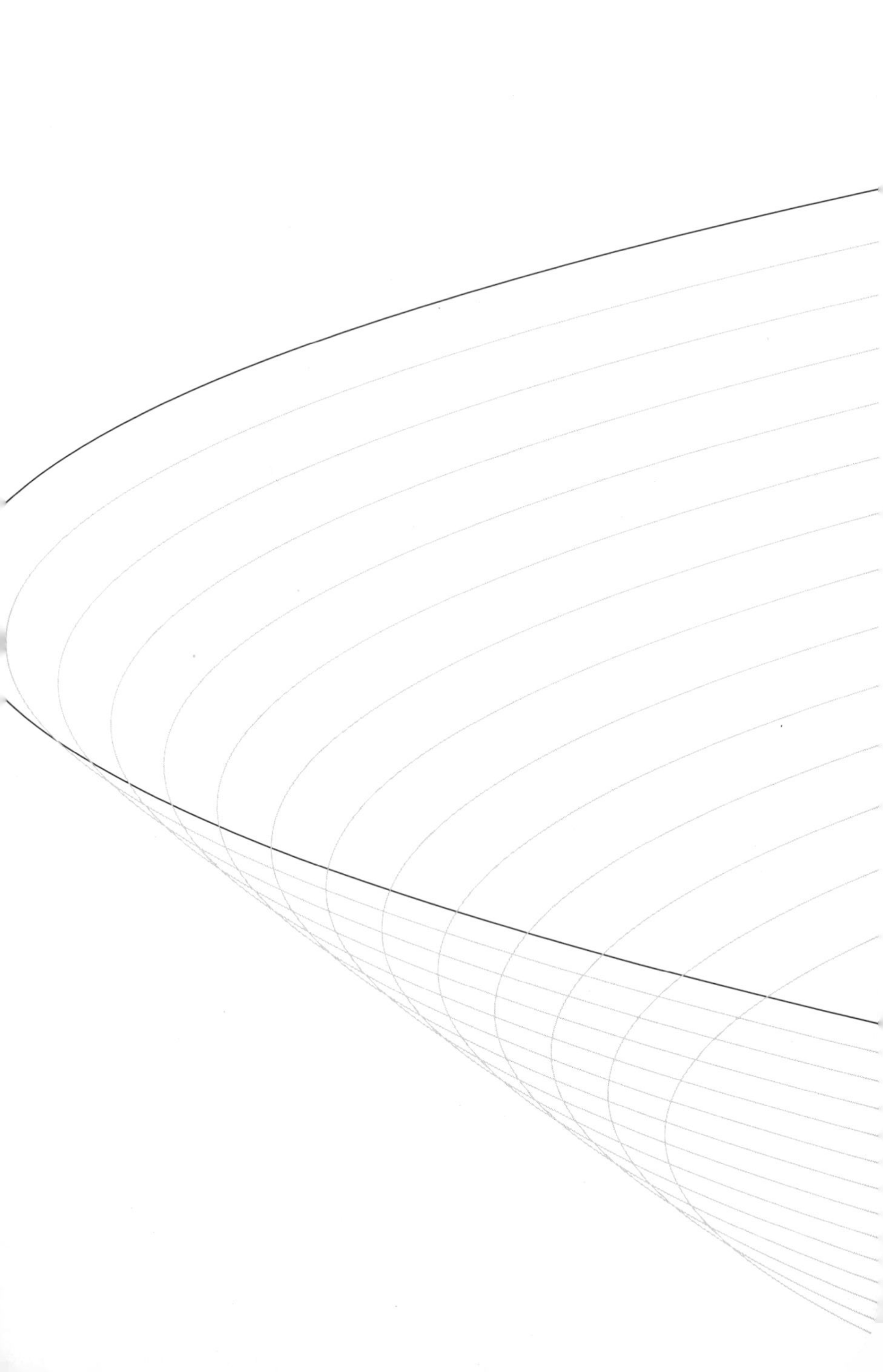

curtain

$y = -\sec^2(c)\,x + \tan(c) - c\sec^2(c) + 3$,
shifted to the left

potbelly $y = c/2$; when a line intersects the circle $\left(x - \frac{5.5}{2}\right)^2 + \left(y - \frac{7.55}{2}\right)^2 = 4$, draw a parabola between the edges of the circle

heart

$y = c$ except when it would intersect $15^2 - \left(|x| - \frac{y}{2}\right)^2 - y^2 = 0$, shifted up and to the left; the interior of the heart is $\cos\left(-2x - 2y - \sqrt{15^2 - \left(|x| - \frac{y}{2}\right)^2 - y^2}\right) = 0$, shifted up and to the left

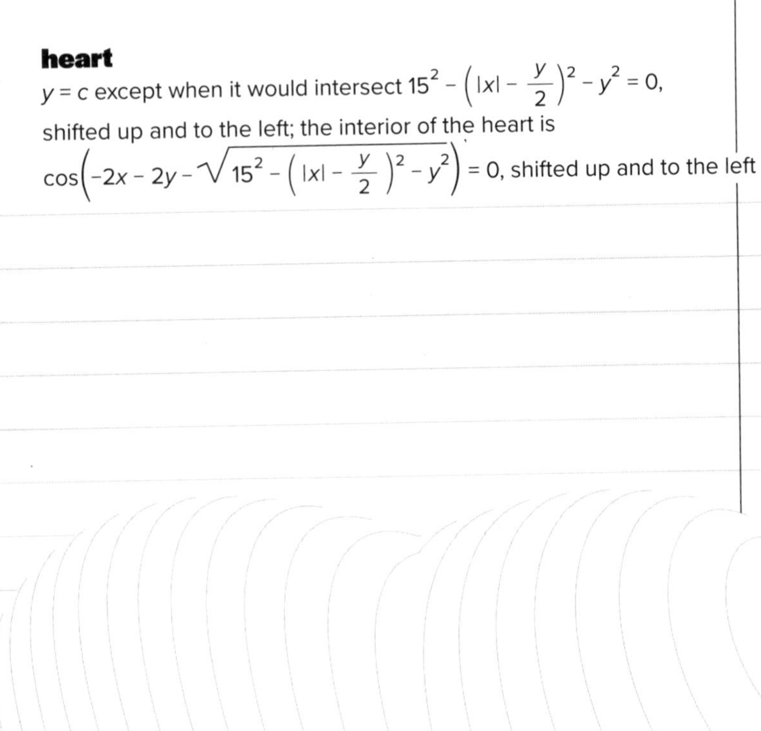

swoop

$$x = 5\left(\frac{-2t\left(t^2-1\right)}{\left(1+t^2\right)\left(1+t^2\right)}\right) + 2.6;\ y = 2.5\left(\frac{-\left(0.1t^2c-1\right)}{\left(t^2+1\right)}\right) + 5.9$$

vortex

shrink black square by 10%; shrink resulting square by 10% and rotate 5.7°, 6 times; then shrink last square by 10% and rotate −5.7°, 6 times

open book

$y = 0.5(x - 2.75)^{2/3} + c$ and $x = 2.75$, rotated 7.5°

corner in middle

line segments connecting the points (0, 3.625), (−2.625, *c*), and (−5.25, 3.625)

collapse

$$y = \ln\left(-\frac{cx}{10} + 5\right) + c$$

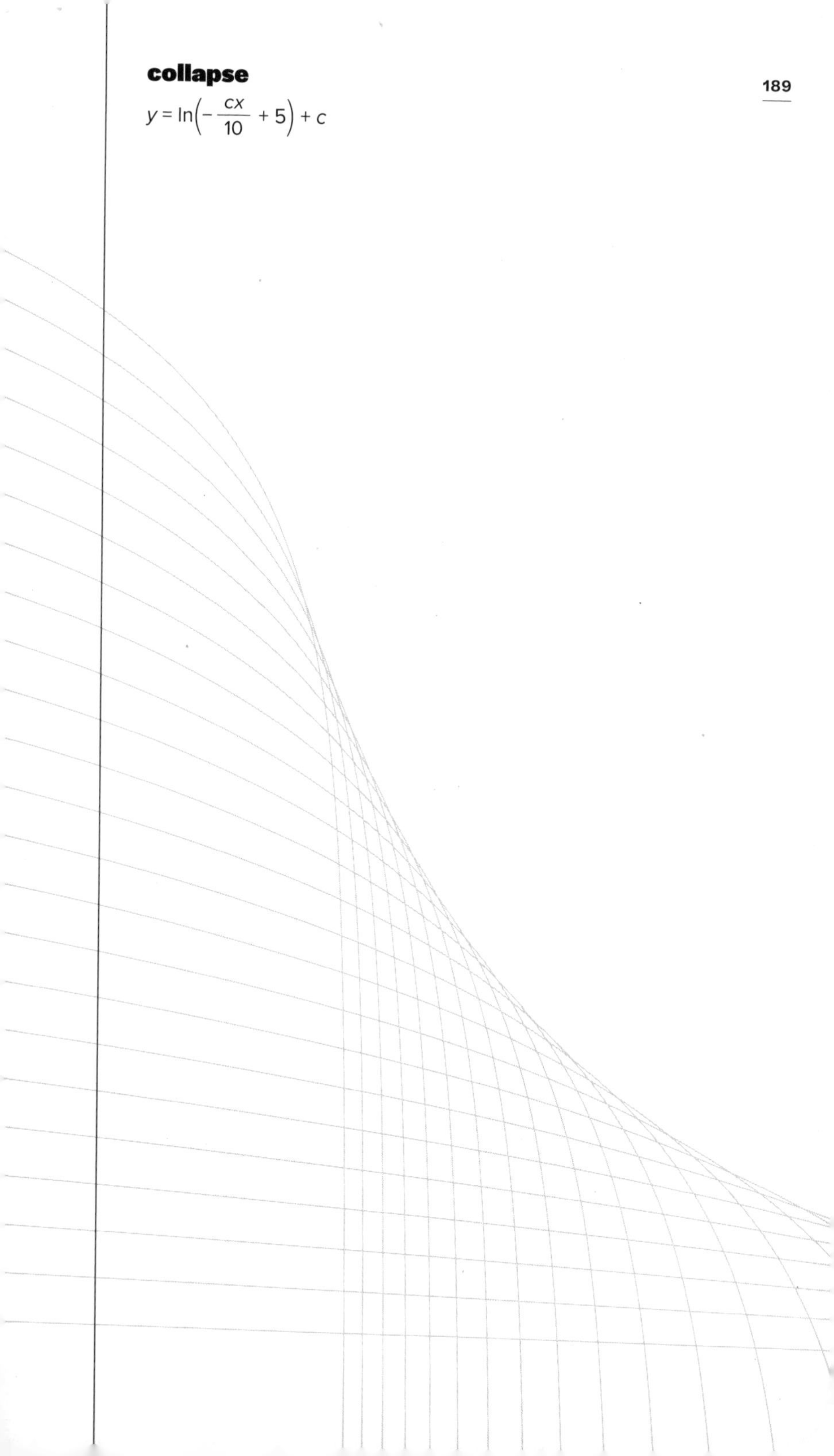

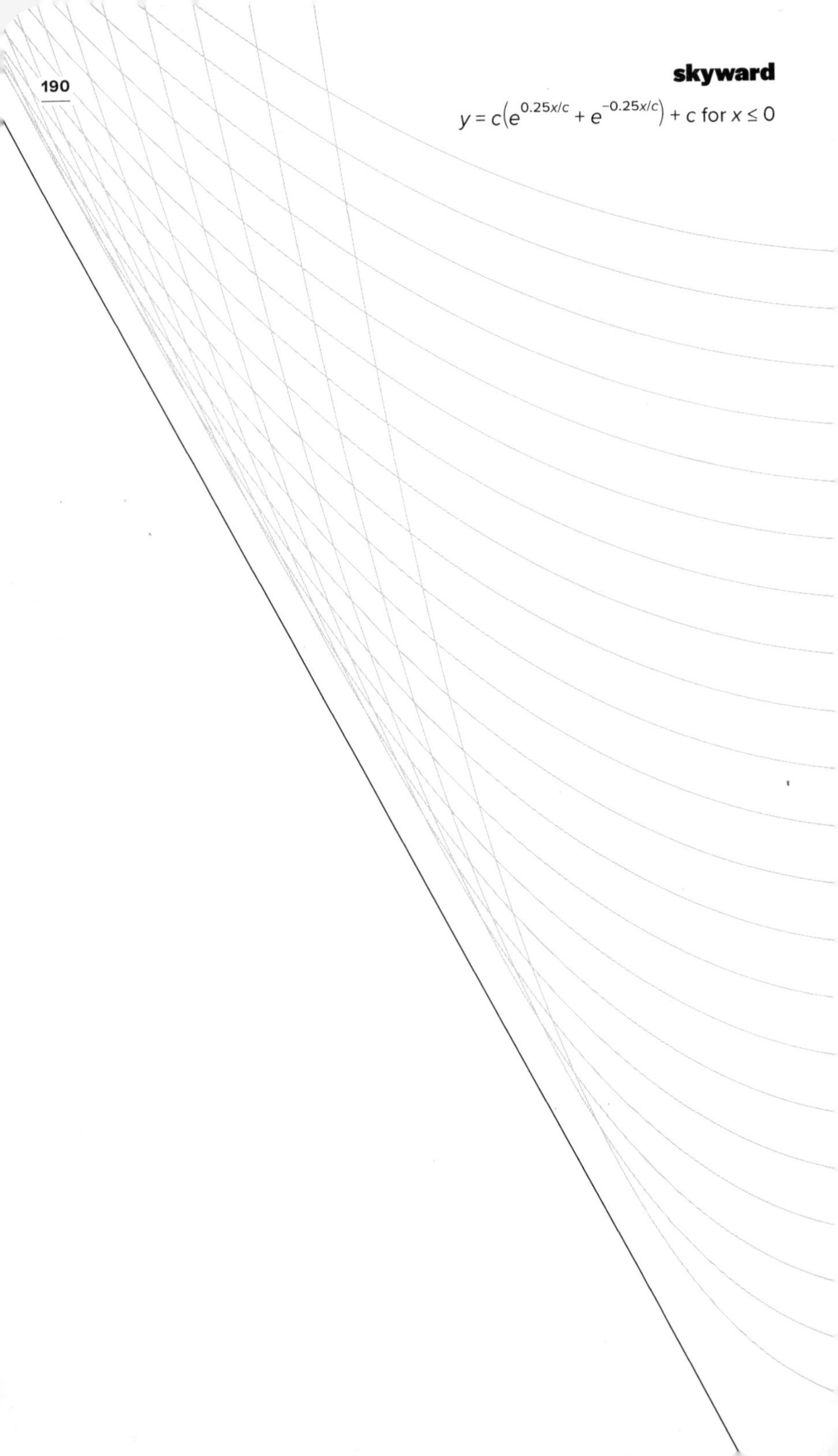
skyward
$y = c\left(e^{0.25x/c} + e^{-0.25x/c}\right) + c$ for $x \leq 0$

cinnamon buns

contour plot of $z = \cos^2(x) + \sin^2(y)$

staircase

the bottom steps are a unit over and a unit down per step;
the top steps are also a unit over and a unit down per step

This 2D drawing of a 3D staircase can be perceived in two ways. First, look at the white region in the lower right and consider this as the closer wall, with the stairs descending from right to left. Second, consider this region as the *farther* wall; the same staircase appears upside down. This optical illusion, known as the Schroeder stairs, demonstrates how 3D perception using 2D data runs into inescapable ambiguities—and how these ambiguities can be a delight to us.

Closing Lines

To assess the quality of an idea, we often invoke "the lines." Dull and conventional? Coloring inside the lines. Subtle and insightful? Reading between the lines. Nuanced and provocative? Blurring the lines.

In this notebook, our aim has been something else: redrawing the lines.

We hope that these nonstandard pages have inspired some nonstandard thoughts. If you put any pages to particularly vivid or exciting uses, we would love to see them. Or perhaps you have been inspired to create your own pages. You can use the free, user-friendly calculator at Desmos to design your own parameterized functions for the 26 notebook-page lines. Or, if you are able to code in Python, there are templates to give you even greater control of the images. To submit a scan or a photograph to us, or to find links to Desmos and Python templates, visit this book's page at https://press.uchicago.edu.

How to Learn More

These pages draw on ideas from across the math curriculum, from Algebra 1 (usually an eighth- or ninth-grade course) to Calculus 3 (usually a second- or third-year undergraduate course). A great tool for further exploration is the graphing system Desmos (https://www.desmos.com/calculator).

YouTube is also an invaluable library. Some channels (such as Khan Academy) offer comprehensive, skills-based sequences of lessons. Others (such as 3Blue1Brown) give beautiful visualizations of university-level concepts. Still others (such as Numberphile) explore fun mathematical ideas from off the beaten path.

There are also shelves' worth of books sharing playful, accessible versions of mathematical ideas. Two of particular relevance to the mathematics here are Paul Lockhart's *Measurement* and Steven Strogatz's *Infinite Powers*.

But of course, there is no substitute for a good teacher. So also worth mentioning is the "Math Help" subreddit (https://www.reddit.com/r/MathHelp/), an online forum where strangers answer one another's math questions.

Acknowledgments

The concept for this book springboarded from Marc Thomasset's Inspiration Pad and tweets from Matt Enlow (@CmonMattTHINK). That work inspired pages like those in this notebook to be created and interspersed throughout the Deconstruct Calculus series by Amy Langville, Kathryn Pedings-Behling, Tyler Perini, and Alex Baham. Put frankly, students loved the pages! So much so that we decided to create a whole book of mathematically inspired nonstandard notebook pages.

We are grateful for innovative ideas on social media. Specifically, Ayliean MacDonald's (@Ayliean) Twitter feed inspired the curve stitching in chapter 9, from which a cardioid emerged. The page titled "wormhole" in chapter 10 was adapted from Joe DiNoto's Twitter feed (@mathteacher1729). The page titled "heart" in chapter 12 was adapted from the Twitter feed @TETH_Main.

And, finally, a huge measure of gratitude to Ben Orlin and our editor, Joe Calamia. Thank you for all you contributed to the book.